FE 20 '87

5-29-97

MW01675419

U.S.A.

DATE DUE

LIVES AT STAKE

Also by Laurence Pringle

SCIENCE FOR SURVIVAL SERIES

Ecology
Energy: Power for People
Nuclear Power: From Physics to Politics
One Earth, Many People: The Challenge of Human Population Growth
The Only Earth We Have
Our Hungry Earth: The World Food Crisis
Pests and People: The Search for Sensible Pest Control
Recycling Resources

EXPLORING AN ECOSYSTEM SERIES

City and Suburb
Estuaries: Where Rivers Meet the Sea
The Gentle Desert
The Hidden World: Life Under a Rock
Into the Woods

6/4/
DRI

LIVES AT STAKE

The Science and Politics of Environmental Health

LAURENCE PRINGLE

MACMILLAN PUBLISHING CO., INC.
New York
COLLIER MACMILLAN PUBLISHERS
London

Macmillan Publishing Co., Inc.
866 Third Avenue, New York, N.Y. 10022
Collier Macmillan Canada, Ltd.

Printed in the United States of America

10 9 8 7 6 5 4 3 2 1

LIBRARY OF CONGRESS CATALOGING IN PUBLICATION DATA
Pringle, Laurence P
Lives at stake : the science and politics of environmental health.
(Science for survival series)
Bibliography: p.
Includes index.
SUMMARY: Presents an overview of environmental factors that affect human health, assesses their dangers, and discusses the scientific and political complexities concerning a healthful environment.
1. Environmental health–United States–Juvenile literature. 2. Environmental policy–United States–Juvenile literature. 3. Medical policy–United States–Juvenile literature. [1. Environmental health. 2. Environmental policy. 3. Medical policy] I. Title.
[DNLM: 1. Environmental health. 2. Politics. 3. Science. WA30 P957L]
RA566.3.P75 363.1'7'0973 80-14272
ISNB 0-02-775410-3

CONTENTS

CHAPTER ONE

Hazard of the Week

Late in the 1970s, a cancer scare struck the town of Rutherford, New Jersey. Health officials discovered that unusual numbers of people had leukemia (cancer of the blood) or Hodgkin's disease (cancer of the lymphatic system). Six of the thirteen leukemia victims were young—between the ages of five and nineteen. For this age group, Rutherford's leukemia rate was nine times the national average.

Panic spread through the town when people discovered that several of these young victims had attended the same elementary school. Some parents wanted the school closed; a few took their children out of school. Health officials swarmed through the community, testing its water and air, and even checking the school's television sets for excessive radiation.

The cause of the cancers remained a mystery. There *were* detectable amounts of nine cancer-causing substances in Rutherford's air, but the same can be said for many other

Site of the leukemia "cluster" in Rutherford, New Jersey

parts of heavily-industrialized New Jersey. Some health experts thought that the cancer "cluster" was just a coincidence.

Coincidence or not, the citizens of Rutherford did not feel as safe as they once had in their town, their schools, and their homes. A nagging fear remained: There might be *something* in their everyday environment that would strike again, and more people, especially young people, would die.

People in the United States are becoming more worried about the healthfulness of their environment. This may seem bizarre to people in some other nations, where lives are often threatened by infectious diseases and where children often fail to survive a year of life. By almost any measure, the United States' environment is among the most healthy on earth. Nevertheless, people worry. They are concerned about real health hazards, some of which only exist in rich, industrialized nations.

In its broadest sense, the word environment refers to everything and everyone around us. It can be applied to all sorts of health hazards, not just in the air we breathe but in our food and in psychological stress at home or work. Environmental health hazards can include a disease-carrying housefly, a reckless driver, and the emotional environment of one's family. Thus in their assessment of cancer-causes, for example, medical authorities include the effects of smoking and diet, as well as of air pollutants.

The general public, however, often thinks of environmental health in a narrower sense, mostly as harmful substances that we have made and put into our air, water, food, and even into medicines and drugs. At times it does seem that we are poisoning ourselves with our own products and processes. Besides cancer, a variety of other illnesses, birth defects, and even behavioral problems of children have been traced to human-made substances used in everyday life.

People have reason to worry, but just how serious is the problem? Can we trust the air we breathe, the water we drink, the food we eat? Must we move to the country and raise our own vegetables and other foods in order to be safe? Or is much of this concern an overreaction to some inevitable risks of modern life?

This book attempts to answer these questions. It presents an overview of environmental effects on our health, assesses the dangers, and describes the complex scientific and political factors which contribute to the uncertainties surrounding this vital issue.

Just a few decades ago, people had a growing trust in the healthfulness of their environment. There was a sense that we could control our world and solve just about any problem. Technological progress was an unquestioned good. After all, so-called miracle drugs, like penicillin, had helped prevent so much infection. Beginning in 1954, mass immunization brought under control the disease poliomyelitis (infantile paralysis) which had killed and crippled many young peoplc.

There seemed to be extraordinary progress on other fronts as well. An insect poison named DDT was introduced. It proved to be very effective and in tropical countries helped reduce the threat of malaria and other insect-borne diseases. Nuclear energy was being developed, and the government agency then in charge, the Atomic Energy Commission, assured people that it would be clean and safe. All sorts of new chemical compounds were being manufactured and introduced into the environment. One company's advertising slogan—"better living through chemistry"—seemed to express how people felt about all technological change.

This optimism began to fade in the late 1950s. Perhaps the first jolt of reality was brought by thalidomide, a sleep-inducing drug that was prescribed to pregnant women by doctors in West Germany and several other nations. In 1959, thalidomide was found to be a *teratogen,* a substance which causes malformations of babies as they develop within their mothers. The unfortunate victims received a great deal of publicity, and so did the then-unfamiliar image of doctors prescribing a sort of poison. Later, it was learned that approval of thalidomide use in the United States had been blocked only because of the strong efforts of one woman. Dr. Frances Kelsey, a physician and pharmacologist for the Food and Drug Administration (FDA), refused to approve thalidomide despite great pressure from the company that hoped to sell the drug in the United States.

During the early 1960s, it also became clear that Americans' beloved automobiles were sources of harmful air pollutants. Then the 1962 publication of Rachel Carson's *Silent Spring* brought the first of a lot of bad news about the side effects and dangers of DDT and other long-lasting insect poisons.

The years since have produced a steady flow of incidents and revelations—of "safe" products that weren't, of drugs that had serious, unforeseen side effects, and of health risks experienced by millions of workers on their jobs. People keep finding more and more worm holes in the fruits of technology. At times, it seems there is a health hazard of the week announced on television, in newspapers, and in other news media.

People relieve their anxiety by joking, "Soon, everything we eat or drink will cause cancer." Since some environmental health hazards, especially cancers, may take decades to develop, people can also ease their worry by saying, "It won't happen to me."

Dr. Frances Kelsey of the Food and Drug Administration

MAULDIN'S VIEW

'No doubt about it. Living causes cancer'

People also try to relieve anxiety by criticizing the source of the bad news—scientists, their experiments, and government regulatory agencies. They sense, correctly, that there is often disagreement among scientists about health risks in the environment. If experts disagree, what are ordinary citizens supposed to do? The national mood seems to be one of confusion and apathy. It is exploited by those who stand to gain by maintaining the status quo. If change is going to occur, people must learn more about the science and the politics of environmental health.

Health and politics? This may seem to be a strange combination, but politics plays an important role in many aspects of life, including health. Almost every day, lobbyists representing industry and others representing the public try to influence the administration, senators, congressmen, and the staffs of government agencies on health issues. The resulting political decisions, both today's and those stretching back for decades, affect everybody's health.

Politicians tend to follow public opinion rather than lead it, so progress in reducing hazards in our environment depends ultimately on public perception of the problem. The pace of change will quicken when people realize that continued discoveries of new health hazards are not inevitable. We already have the scientific and technological knowledge needed to prevent many existing environmental health risks. Whether we choose to use this knowledge depends largely on how well people understand the science and the politics of this crucial issue.

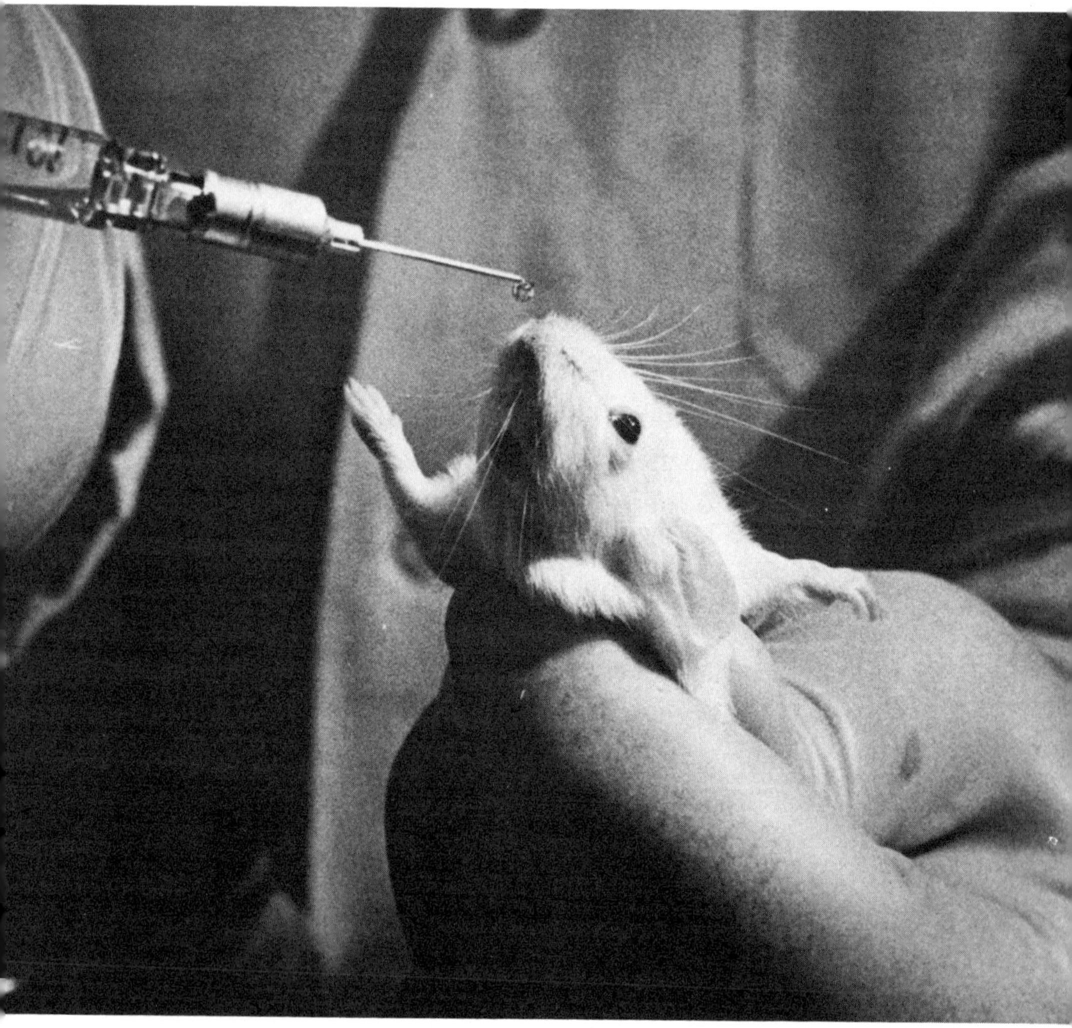

CHAPTER TWO

In Scientists We Trust

Opinion polls reveal that people tend to trust scientists much more than they do businessmen or politicians. In matters of environmental health, however, many people feel annoyed and frustrated by the actions of scientists.

Most laymen believe, mistakenly, that scientific fact is absolute, with no uncertainty. But in science and every other human endeavor, absolute knowledge is rare. Scientists sometimes disagree and often admit that they can't give black and white answers. About several environmental health problems they have said, "On the one hand, the evidence is such-and-such, but on the other hand. . . ."

This prompted one United States senator to plead for more "one-handed" scientists. Like many other people, the senator wanted scientists to provide easy solutions to difficult problems. This is unrealistic, wishful thinking. Present knowledge about many aspects of environmental health

Mice, rats, and other small mammals are vitally important in scientific tests of both harmful and benefical substances.

leaves plenty of room for doubt. This scientific uncertainty and the credibility of scientists are central issues in controversies over environmental health. Consider the case of saccharin.

First sold in 1900, saccharin gradually became a widely-used, calorie-free sugar substitute. Its use in low-calorie soft drinks, chewing gum, and many other products grew enormously after 1969, the year the Food and Drug Administration banned another noncaloric sweetener, cyclamate. Tests with laboratory animals indicated that cyclamate could cause cancer of the bladder. Although seven out of ten families in the United States used cyclamate at the time, the banning caused no great protest because people knew that there was a substitute, saccharin.

Since 1912, saccharin has passed several safety tests. (Through hindsight, however, scientists now see flaws in those tests.) In the 1920s it was placed on a list of "generally recognized as safe" (GRAS) food additives by the FDA. In 1972, however, a new study suggested that saccharin could cause cancer. Its GRAS-status was revoked. The National Academy of Sciences (NAS) was asked to consider all of the studies completed up to that time; the academy said that the evidence about saccharin's safety was inconclusive. In 1977, however, a Canadian study showed that saccharin caused bladder cancer in rats. The Canadian government immediately banned the general use of saccharin. The FDA announced its intention to do the same in the United States.

This announcement brought a storm of public protest, and intense lobbying by such special interest groups as beverage companies and the Calorie Control Council, which is

The FDA required a warning label on many products containing saccharin.

mostly funded by the saccharin and diet food and drink industry. The FDA brought some of this trouble on itself. In an effort to allay public alarm about the danger of saccharin, its announcement mentioned that the rats used in the study had been given the equivalent in saccharin of 800 cans of diet soda a day. This use of a heavy dose was a sound procedure, endorsed by more than twenty scientific reports since 1956, but it loosed a flood of editorial ridicule and mocking cartoons. A congressman suggested that saccharin products carry this warning: "The Canadians have determined that saccharin is dangerous to your rats' health."

The situation in Canada was markedly different from that in the United States. The Canadian Diabetes Association supported the government ban on general sales of saccharin, admitted responsibility for promoting overuse of saccharin, and actively helped reverse that teaching.

In the United States, a representative of the American Diabetic Association argued that diabetics, for whom sugar can be lethal, need "soda without fear." Doctors testified that they believed saccharin helped people control weight. And a spokesman for a toothpaste company warned that removal of the sweet taste of saccharin from toothpaste would cause a major increase in dental diseases.

Congress allowed continued use of saccharin for eighteen months (later extended) while further studies were made, and a search for a saccharin replacement was speeded up. The NAS was told to review all available evidence and make recommendations. A committee of scientists made the first part of the NAS report in late 1978. They concluded that saccharin poses a cancer risk in humans, though probably a low risk. The committee said that the cancer risk is well established, and that no further tests are needed on this issue. The scientists also reported evidence that saccharin is a "promoter" of cancers caused by other substances.

According to this NAS committee, too little was known to be able to predict the number of human cancers that would be caused by saccharin consumption. They warned that two groups seemed especially vulnerable: women of childbearing age and children up to ten years of age. Between 1972 and 1978, saccharin use by young children increased 160 percent. The amount they consume, relative to body weight, gives children a much greater exposure than adults to the cancer risk of saccharin. Cancers usually develop slowly, so the full effect of this saccharin use by children won't be known for many years.

Among other unknowns is whether saccharin has any health benefits. Of course, many people believe it does, but belief is not proof. The NAS committee concluded that attempts should be made to either confirm or disprove this belief.

Actually, the alleged benefits of saccharin have been in dispute for some time. Studies show that saccharin causes a lessening of sugar in the bloodstream. A decline of blood sugar is a trigger for appetite, and animal studies in laboratories show that animals that drink so-called diet soda replace the "missing" calories in their drinks by eating more food. Thus the available evidence indicates that saccharin makes it more difficult, not less, for a person to lose weight. Nationally, the annual use of about 7 million pounds of saccharin has had no positive effect on the average American, who has gained weight while consuming so-called diet foods and drinks.

Saccharin's only benefits may be psychological. Besides a sweet taste, it may give dieters and diabetics a sense that they are doing something about their problem, even though they may not be. Being upset about the possible loss of saccharin is understandable. But since the issue has raised all sorts of questions, it seems wise to ask where the public got the notion that saccharin is beneficial. Directly or indirectly, the idea can be traced to advertisements and other promotional activities of the companies that make, sell, distribute, or otherwise profit from saccharin's use. In the three months immediately following thc FDA's proposal to ban saccharin, the Calorie Control Council spent $890,000 on advertising and lobbying. The fact that saccharin may not help people lose weight and look slim is not advertised, while these alleged benefits have been advertised for decades.

In Canada, saccharin is available in drugstores as a nonprescription item. Canadian soft drink manufacturers now market sugar-free drinks to which saccharin can be added. (When consumers add their own, it is usually one or two 15-milligram tablets; commercial diet soda in the United States contains 120 milligrams.) Canadian beverage companies also offer "low-sugar" soda, with half the calories of their regular soda. Soft drink sales dropped after the saccharin ban took effect, but have since recovered. Consumers in Canada now have a wider choice of beverages than before. And according to Canadian medical authorities, the saccharin ban has had no impact on the health of diabetics.

In the United States, the second part of the National Academy of Sciences report was released in 1979. The panel was made up of thirty-seven scientists, attorneys, and public policy experts. This group seemed more convinced of saccharin's harmfulness, but made no specific recommendations about how to protect the public, except to urge that the FDA be given more flexible general rules, for example, permission to restrict the sales of a product instead of banning it totally.

The tone of the final report was calm and bland, but it was actually a result of bitter disputes. Almost a third of the

Restricted use of saccharin in Canada led to marketing of soft drinks unavailable in the United States, where advertisements (far right) claim that saccharin aids weight loss.

Put less sugar
in your life.
Get more out of living!
Wherever you go, Sweet 'N Low brand Granulated Sugar Substitute helps keep families looking right. Feeling light. Used as a sweetener instead of sugar, or for cooking and baking, Sweet 'N Low can help you stay trim. And your taste can't tell the difference. Try it. Be in the know.
Switch to
SWEET'N LOW
Granulated Sugar Substitute
The Great
Performer

Diet
TaB
TRADE MARK
MARQUE DE COMMERCE
calorie-reduced Diet Tab réduit en ca
TRADE MARK
MARQUE DE COMMERCE
Diet
TaB
TRADE MARK
MARQUE DE COMMERCE

panel members disagreed strongly with some of the report's recommendations. For one thing, this minority urged a total saccharin ban.

The major conflict was not about saccharin, however. That was a jumping-off point for the main focus of the report—criticism of existing food safety laws and suggestions for sweeping change. Current food regulations were called "complicated, inflexible, and inconsistent." There is certainly some truth in this; the laws which the FDA administers have been amended and patched together by Congress through the years.

A majority of the NAS panel recommended that the food safety laws be simplified and that the FDA be allowed to consider benefits as well as risks of additives. Food substances could be assigned to one of three risk categories—low, medium, or high—and the FDA could take a variety of actions for each category. For a food additive with a proven risk of causing illness, for example, the FDA could ban its sale outright, restrict its sale, or require a warning label stating the risk clearly.

Dissenting members of the panel were especially critical of this idea. They maintained that there is currently no "scientifically defensible" way to divide unhealthy substances into different risk categories. Their statement said, "The ability of science to quantify human risk has not advanced sufficiently . . . to permit the construction of a scientific rationale for such a scheme."

To attempt to do so, they argued, would open the FDA to even greater political pressure from industry and would lead to less protection for the public.

Much of the debate about food safety law centered on the

Delaney clause, an amendment added to federal food safety laws in 1958. It prohibits addition of *any* amount of a chemical to food or drugs once it has been found to cause cancer in humans or other animals. The Delaney clause obligates the FDA to outlaw even tiny amounts of a cancer-causing substance (carcinogen), if presence of the substance is avoidable.

Critics of this clause contend that it is too strict, partly because of changes in chemical analysis that have occurred since 1958. At that time, 50 parts per million (ppm) was about the smallest amount of a contaminant that could be detected. Now, levels as low as 10 parts per trillion (ppt) can be measured.

The Delaney clause was applied in the case of beverage bottles made of acrylonitrile (AN) plastic. Tests showed that a beverage left in an AN bottle for six months became contaminated with an average of 10 parts per billion (ppb) of AN. In a legal sense this made AN plastic a food additive. Since it was shown to cause cancer in laboratory animals, the FDA banned the use of this kind of plastic bottle.

The manufacturer argued that the FDA was unnecessarily strict and that the tiny amounts of AN found in beverages would be harmless to people. But is any amount of a substance that causes cancer safe? Among scientists, the consensus is that there is no proof that a safe level, or threshold, exists for any carcinogen.

This was true in 1958, when the Delaney clause became law. Since then, no progress has been made in determining a safe amount of a carcinogen. In fact, new evidence suggests that estimating a safe dose is more difficult than previously thought.

Tests with laboratory animals show that cancers can be induced by very low doses of some carcinogens. The age of animals can be a vital factor, too. In one study, animals exposed to the plastic vinyl chloride were more likely to get cancer if they were newborn when exposed. And we now know that some substances seem to stimulate or promote other chemicals to be carcinogens. This is a synergistic effect, in which the effect of two or more chemicals acting together is greater than their separate effects.

Information like this makes it more difficult than before to say that a certain amount of a chemical is safe, or to estimate the risk to the human population. Because of this, some health experts were alarmed when the furor over saccharin seemed to threaten the Delaney clause. Legally, the FDA had good reason to ban saccharin in 1977, and the scientific evidence against saccharin has since grown stronger. (Saccharin has caused cancer in rats at doses as low as the equivalent of just 1.6 bottles of diet soda a day.) As former FDA Commissioner Donald Kennedy pointed out, "The general provisions of the food safety laws would have required a move to ban saccharin. If you want to save saccharin, you're going to have to change more in the law than the Delaney Amendment."

With the controversial NAS report as a stimulus, Congress began to consider basic changes in federal food-safety laws. A major battle loomed, with industry lobbying for weaker regulations and consumer groups lobbying for strict prohibition of unsafe substances in food. As the furor increased, an aide to a United States senator complained, "We have to contend with a well-financed industry lobby against a divided scientific community."

In general, industry cannot be counted on for compassion in matters of environmental health. With few exceptions, industry regards possible health hazards primarily as *business* problems. Having spent many thousands and perhaps millions of dollars to develop, produce, market, and advertise a product, manufacturers tend to view criticism of that product as a threat to their business health. They spend extraordinary sums to influence members of Congress, other politicians, and the staffs of such regulatory agencies as the FDA. And when a product is proved to be a health hazard and is banned or restricted in the United States, its manufacturer often continues to sell it abroad, neglecting to tell its new customers about the proven hazards.

It seems clear that decisions about people's health should not be entrusted to industry lobbyists. The general public played into the hands of these powerful forces when it protested the FDA action on saccharin.

There would have been much less public opposition to a saccharin ban if a ready substitute were available. But cyclamates had already been outlawed (on the basis of less firm evidence than that against saccharin), and no other noncaloric sweetener was approved for widespread use. One candidate, Aspartame, even if approved, lacks some of saccharin's qualities. It eventually breaks down in water or from exposure to high temperatures. Aspartame and other sweeteners, such as xylitol and sorbitol, are under investigation for suspected ill effects on people.

Considering the FDA's belated discoveries about the dangers of saccharin and the fact that an approved artificial sweetener would be used daily by as many as 70 million people in the United States, it is understandable that the FDA

Aspartame was evaluated as a saccharin substitute.

would be cautious in evaluating a substance which might eventually take saccharin's place.

In 1980, the saccharin saga continued with reports of three new studies. Newspaper headlines read "Three Saccharin Studies Find No Cancer Link" and "Saccharin Scare Debunked." The diet food and beverage industry claimed that saccharin had been declared safe, and Congress took pride in its 1977 decision to overrule the FDA.

Unfortunately this was just further evidence that most journalists and congressmen are scientific illiterates. The new findings were based on epidemiological studies, which involve comparisons between the life styles of disease victims and those of other people in the same population. The most elaborate study, conducted by the National Cancer Institute (NCI), surveyed 3,000 bladder cancer patients and 6,000 healthy individuals of the same ages. This was too small a sample to detect whether saccharin is a weak carcinogen. Officials of the NCI and the FDA knew this beforehand. The NCI study was aimed at finding whether saccharin caused a higher cancer rate than expected from studies with laboratory mammals. It did not. Dr. Robert Hoover of the NCI concluded that "nothing in this study, or the other two, is inconsistent with saccharin being a weak carcinogen."

Dr. Hoover also warned that "any use by nondiabetic children or pregnant women . . . and excessive use by anyone is ill-advised and should be actively discouraged by the medical community."

The consensus among scientists, that saccharin poses a low but real risk of cancer to people, was not changed by the new studies. However, misunderstanding about saccharin ran deep among the public and in Congress, and the FDA seemed resigned to allow widespread use of saccharin to continue.

The saccharin case is just one of dozens of environmental health issues, and it is not even among the most serious. Nevertheless, it shows how complicated and difficult it can be to ensure that we have safer food and a safer environment. It also shows that progress toward that goal can be blocked by an ill-informed citizenry which cannot distinguish between promotional claims and real benefits. Progress toward a healthier environment depends more on a better understanding of the issues than on new medicines or new laws.

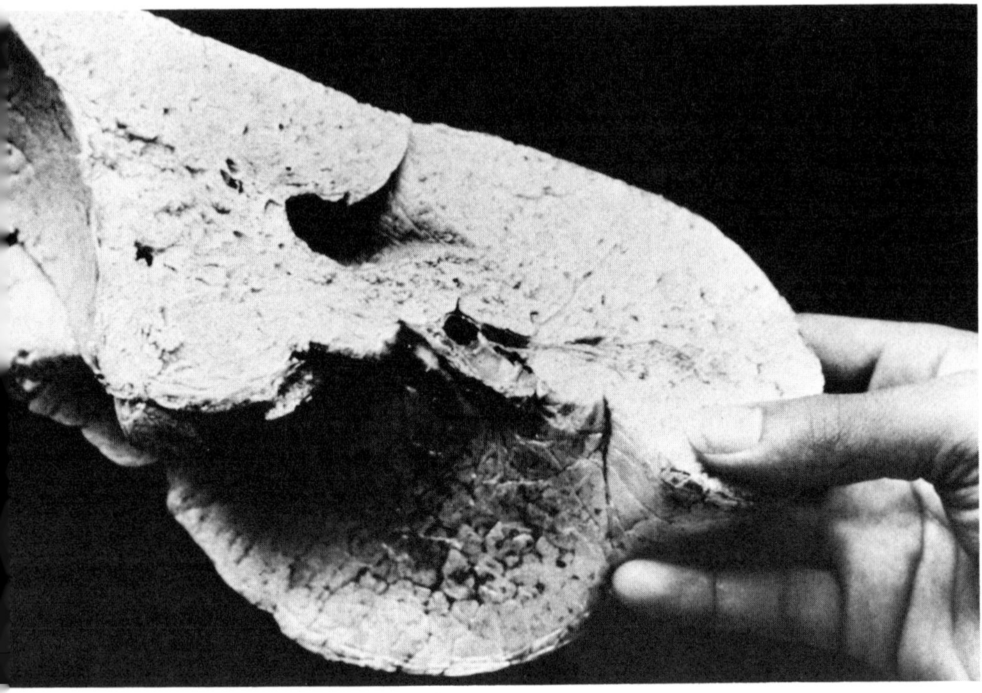

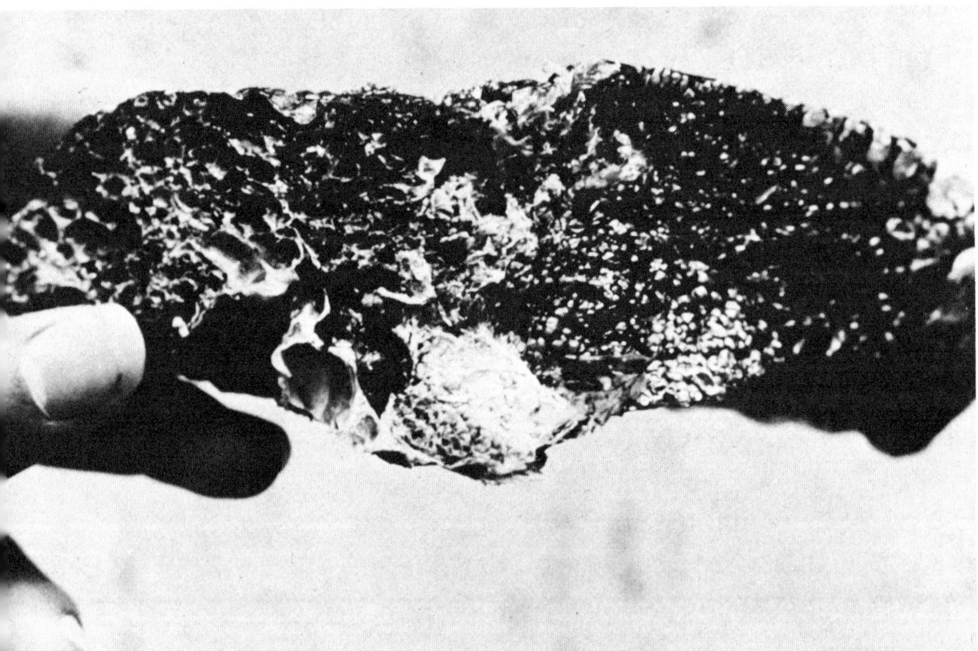

CHAPTER THREE

Testing for Health Hazards

Some diseases have rather obvious causes. An autopsy shows the accumulated coal dust that gave a miner black lung disease. Other cause-effect connections are less obvious. For example, why do truck drivers have a greater than normal incidence of skin cancer on their left forearms? This puzzled physicians for a while. The cancers are a result of the drivers leaning their left arms out of their rig's open window. The left arm is exposed to many more ultraviolet rays than the rest of the body.

For some other health hazards though, it is difficult to establish a clear, one-to-one relationship between a disease and its cause. Sometimes a disease-causing substance must accumulate in a person for many years before its total quantity causes illness. Cancers have a long latent period—the time between exposure to a carcinogen and the noticeable development of malignant tissues.

A normal adult lung (above) compared with the lung of a coal miner (below).

Even so, it is sometimes possible to find the cause of a disease by studying not just a single case, but a whole group of people. Epidemiological studies were of no value in the "cancer cluster" case in Rutherford, New Jersey, because the life styles of the victims seemed indistinguishable from those of other people in the community. But epidemiology has contributed to the evidence linking lung cancer to cigarette smoke. One factor—smoking—distinguishes some people from the rest of a population. By comparing the health of smokers and nonsmokers, a clear connection between smoking and rates of lung cancer can be seen.

Epidemiological studies have also produced strong evidence about health hazards in places of work. In 1977, about half of the male workers in a California chemical plant were found to be sterile. They all had one thing in common: They had handled a pesticide called DBCP. And in 1978, the Department of Health, Education, and Welfare (HEW), which in 1980 became the Department of Health and Human Services, announced that nearly half of the 8 to 11 million people who had worked with asbestos during the past few decades may die of cancer or of asbestosis, also known as white lung disease.

As time goes by, new epidemiological evidence will help identify other harmful substances in our environment. This is one way of learning about health hazards, but it is an inhumane and tragic one. Waiting for disease and death to affect people—the body count method—will not do.

Using humans as test animals is unnecessary, because decades of research have shown that other animals can be used to test chemicals, with the results giving a good indication of effects on people too. This fact has long been accepted by medical and scientific experts, so they were surprised and alarmed over the public reaction to the Canadian animal tests which indicated that saccharin was a carcinogen. The jokes and scorn about saccharin experiments revealed a disturbing lack of understanding of methods long trusted by scientists.

Gus Speth, chairman of the President's Council on Environmental Quality, said, "The ridicule arising from the saccharin episode may stimulate contempt about our regulations of other widely used but potentially hazardous chemicals—and we may find ourselves, some years from now, laughing all the way to the hospital."

Here are two common complaints about animal tests of chemicals and responses to them:

People aren't rats. Just because a chemical causes cancer or some other sickness in rats doesn't mean it is harmful to humans.

In 1978, the Congressional Office of Technology Assessment declared that animal testing provides "valid, reliable prediction that a substance will produce cancer in humans." The same is true of tests for most other health hazards; there is wide agreement among scientists about this.

We are quite lucky that such fellow mammals as rats, mice, other rodents, and sometimes rabbits and monkeys are reliable subjects for such tests. They may seem very different from people, but on the cellular and molecular level, where cancer and many other diseases take effect, the differences are minor.

Many substances were first shown to be carcinogenic in small mammals, then later proved to cause cancer in people. And cancers often form in the same body organs of both humans and small mammals. In this and other ways, rats and mice are good "stand-ins" for people.

There are differences, of course, not just between people and small mammals but also between different kinds of small mammals. Scientists have paid a great deal of attention to these differing characteristics in order to run trustworthy tests on chemicals. For example, one hydrocarbon compound causes skin cancers in mice and rats but not in guinea pigs. A hair-dye ingredient causes thyroid cancer in rats at both high and low doses, but in mice only at high doses. In some instances, humans are more sensitive to the ill effects of a chemical than rats, mice, or other small mammals. Ideally, differences like these must be taken into account when planning and evaluating experiments. Overall, however, a chemical that causes cancer or other illness in one kind of mammal is likely to do so in most others. So far, arsenic is the only substance known to be carcinogenic in humans but not in other mammals.

What about huge doses? Won't large amounts of any chemical cause sickness, including cancer, in test animals?

Large amounts of substances, including table salt and vitamins, can cause illness and death, but no amount, large or small, of a noncarcinogenic material will cause cancer. The ability to cause cancer seems to be rather rare among the many thousands of chemical substances in our environment.

Testing at high doses does not produce false results and make a safe substance appear to be a cancer-causing one. And there are sound reasons for giving test mammals high doses of the substances under investigation. Compared to humans, rodents lead very short lives. Animals that normally live about two years are standing in for people, so tests must exaggerate the effect of a chemical. Large doses accomplish this. Also, if a substance is harmful, the higher the dose, the earlier its effects appear and the earlier its dangers are known.

The amount of a chemical fed to test mammals is also determined by the number of animals used in the experiment. If an ideal experiment could be devised, the number of rats or mice would equal the human population of the United States, about 220 million. That way, everybody would have an experimental stand-in rodent. Actually, they would have two, because the experimental rats would later be compared with 220 million "control" rats of the same

Rodents like these have proved to be good "stand-ins" for humans in tests of many substances.

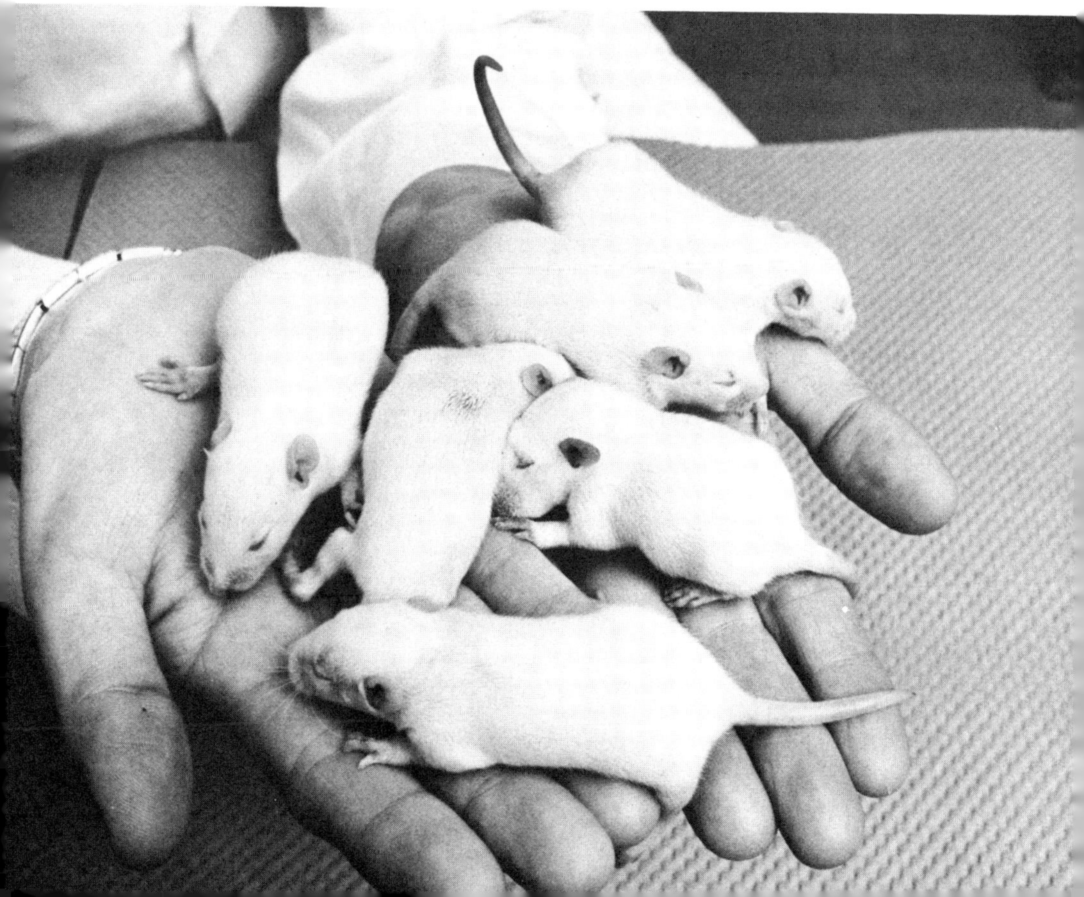

type, which would be fed and cared for in the same way, except they would not be exposed to the chemical being tested.

At the end of a year or two, all 440 million rats would be killed and carefully examined for cancers. An unknown percentage of both groups would inevitably have some cancers, due to heredity or other reasons. This is true of all mammal populations. However, if investigators found an excess of cancers in the experimental rats, they could safely conclude that the chemical being tested was a carcinogen.

In this fantasy experiment, a low dose of the chemical could be used. Even if it had a carcinogenic effect on just one tenth of 1 percent of the population, this dose would produce noticeable results: 220,000 victims of cancer. However, an experiment like this will never be tried, and for good reason. Although rats can be fed cheaply, the cost of maintaining 440 million of them for two years and later performing all those autopsies would run into several billion dollars.

Usually, 600 mice or rats are used. Even with this small number, a well-run experiment costs at least $200,000. Half of the rodents are the experiment's controls. The other 300 are often divided into three groups (each with an equal number of males and females). This enables the investigators to test a chemical at three different doses. But very low doses are not usually used because their effects, if any, might not show up in a population of just a hundred rodents.

Tests like this are not only reliable for detecting whether a substance is carcinogenic but have long been used to determine the effects of drugs, vitamins, diets, and pesticides. Usually the same substance or diet is tested more than once by other investigators, because some experiments may be poorly designed or poorly carried out and may produce false negative or false positive results.

For many years the FDA and other agencies naïvely assumed that they were receiving accurate accounts of well-controlled experiments from industry. Then, purely by chance, their naïveté was lost. In 1972, Adrian Gross, an FDA pathologist, saw a scientific report about positive cancer tests performed on a drug which had been approved by the FDA. He then reviewed the data which had been submitted by the manufacturer and notified the company of some discrepancies in their report. Eventually, the firm sent the FDA a new version of the same study. Gross was startled to discover that, "Instead of changing the summary to more accurately reflect the data . . . the data had been changed to more accurately reflect the summary."

Later, an investigation uncovered evidence of sloppy procedures and even fraudulent test results that had been submitted to the FDA, the Environmental Protection Agency (EPA), and other regulatory agencies. Disturbing numbers of the studies conducted by or for industry were found to be flawed. For example, the Allied Chemical Company hired Hazelton Laboratories to test a food additive (Red Dye Number 40) to see whether it was toxic or carcinogenic. The laboratory concluded that the dye was safe. Allied Chemical submitted the test results to the FDA and made plans to begin selling the dye.

The FDA discovered, however, that the laboratory had not even conducted the usual carcinogen test with mice, and their experiment with rats was practically worthless. Most of the rats died of an infection early in the test. With less than the normal number alive for the whole test, only an extremely strong carcinogenic effect could have been detected.

The results of some carcinogenic tests on pesticides aldrin, dieldrin, chlordane, and heptachlor were also found to be worthless. The manufacturers claimed that tests with mice showed no carcinogenic effect. True enough, but only because high doses of the pesticides killed most of the mice early in the experiments, before they had time to develop cancer.

Industries sometimes try to "explain away" test results. In congressional hearings about certain pesticides, industry spokesmen argued that harmless or benign tumors, not cancers, were found in the livers of the test mammals. But most cancer experts question whether there is truly a "benign" tumor. Tumors sometimes become cancerous. A panel of scientists said that it was "unaware of the existence of any chemical which is capable of producing benign tumors only."

FDA investigators found evidence that other firms used "creative penmanship" which caused test animals to appear and disappear through the course of a study. Some experiments were designed in ways that obscured whatever toxic effects a substance might have. Thousands of experiments by one major laboratory were plainly fraudulent, with glaring differences between test results and the data actually reported. The results had led to FDA approval of nearly a hundred products and to EPA approval of 123 pesticide ingredients.

Exposure of all these deficiencies led to 1976 Senate hearings. New laboratory regulations were proposed by the FDA, and despite opposition from industry and laboratories, these took effect in 1979. The Good Laboratory Practice rules set rigorous standards that must be applied to every study. And Congress gave the FDA enough money to set up an inspection program to enforce the standards.

Besides ensuring more reliable results, the new rules made laboratory tests more expensive—bad news, considering the massive amount of testing needed. More than 50,000 human-made chemicals are produced and used at present. At least 8,000 widely-used chemicals have not been tested for carcinogenicity. And about a thousand new chemicals are put on the market—and into the environment—each year. Only a small fraction of these are tested before their use. We live in a sea of human-made chemicals. Most of them are probably harmless. But it is vital to identify the dangerous substances, and they must be tested one at a time.

"This is no time to be a mouse," observed humorist Russell Baker. However, mice and other rodents may get a break. An urgency to test more chemicals and to do it more quickly has lead to the development of rapid nonanimal tests for some kinds of health hazards. The most promising of these new techniques is called the Ames test. It was developed over the span of twelve years by Dr. Bruce Ames, a professor of biochemistry, and his colleagues at the University of California at Berkeley. It works like this:

Each of several sterilized plates is covered with a food substance and about a billion *Salmonella* bacteria of a particular type. Samples of the chemical to be tested are put on

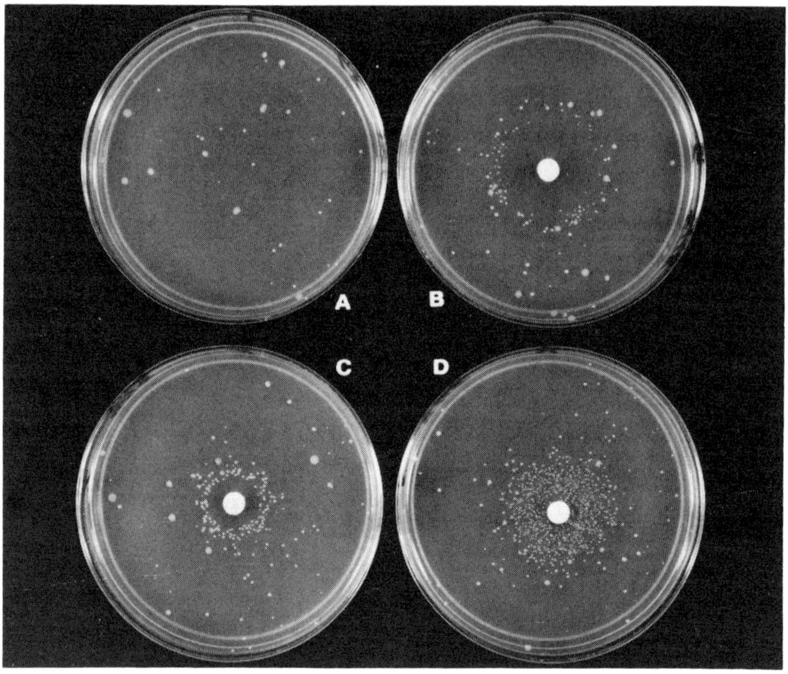

In these examples of Ames test results, the control plate *A* contains colonies of spontaneous bacterial mutations. On plates *B, C,* and *D,* mutagenic substances placed on round paper discs caused additional bacterial colonies to form.

some plates but not on others (the controls). Then all of the plates are covered and for two days are kept at 37° C (99° F), ideal temperature for growth of the bacteria.

Normally, however, this strain of *Salmonella* bacteria cannot grow because a certain food nutrient is missing. They can only grow if some of them change so that they can make the missing nutrient themselves. These changes in the genetic materials of cells are called mutations. Ordinarily, some mutations occur in both the experimental and control groups of bacteria. Each mutated bacterium multiplies and forms a visible colony, a white spot on the plate. After two days the colonies on all plates are counted. If the plates which contained the test chemical have considerably more colonies than the control plates, this means the chemical caused the extra mutations. It is a mutagen of bacteria, and a powerful mutagen produces more colonies than a less potent one.

Mutagenic substances are a major cause of genetic birth defects and may contribute to heart disease. Also, it seems that both mutagens and carcinogens cause damage to genetic material of cells. This suggests that mutagens are likely to be carcinogens. With some exceptions, this has proved to be true. Dr. Ames and his colleagues tested 176 known carcinogens, and about 90 percent were also mutagenic.

The Ames test enables scientists to detect mutagens and potential carcinogens in just three days, at a cost of a few hundred dollars per chemical. No wonder the test is now used in over two thousand laboratories around the world. Among other advantages, it enables industries to weed out hazardous chemicals at early stages of development. One chemical company had a very promising pesticide which seemed to be strongly mutagenic in the Ames test. Chemists investigated further and found that the mutations had been caused by an impurity. With the impurity removed, samples of the pesticide were not mutagenic, so the company decided that it was worthwhile to invest money and time in further experiments with laboratory mammals.

The Ames test has already provided quick warnings of health hazards and brought attention to several cancer-causing substances. One was furylfuramide (AF-2), an antibacterial food additive used widely in Japan. Preliminary tests with mice and rats showed no carcinogenic effect. Then, in the early 1970s, Japanese scientists found AF-2 to be a strong mutagen with *Salmonella* and other bacteria. This lead to more thorough tests with mammals, and the results showed that AF-2 was a carcinogen in rats, mice, and hamsters. The Japanese government prohibited the use of AF-2, and all products containing it were removed from the market. Large amounts had been consumed for nine years, but the effects on the Japanese population won't be known for many years because of the latency period of cancers.

Some ingredients of hair dyes are known to be absorbed through human skin, but few of these chemicals have ever been tested to see if they cause cancer. The Ames test showed that about 90 percent of commercial hydrogen peroxide hair dyes were mutagenic. The National Cancer Institute has since completed mammal tests on several of these chemicals and found them all to be carcinogens.

Information from the Ames test also proved to be vital in the case of Tris-BP, a flame-retardant chemical in children's pajamas. From 1972 to 1977, 50 million children wore sleepwear with Tris added. The chemical can be absorbed through the skin and has, in fact been detected in altered form in the urine of children who wore Tris-treated sleepwear. After Tris was found to be mutagenic, further tests revealed that the chemical caused cancers in mice and rats and sterility in such mammals, too.

Tris was banned from pajamas made or sold in the United States, but for a time, the Consumer Product Safety Commission allowed Tris-treated sleepwear to be exported

A laboratory technician extracts Tris from children's sleepwear.

for use by children in foreign countries. Further exports were barred by the commission only after two new commissioners, appointed by President Carter, helped form a new, more humanitarian majority.

The success of the Ames test has encouraged a search for other short-term ways to detect hazardous substances. Several have been developed, but none has emerged as a substitute for mammal studies. The Ames test itself misses some types of known carcinogens. None of twenty-nine different short-term tests showed chloroform to be carcinogenic, yet mammal tests reveal that it is. It seems likely that in the future, investigators may use several rapid testing methods to complement the still-vital experiments with laboratory mammals.

We are in the midst of a scientific revolution in the ways we identify health hazards. It is a time of change and uncertainty, but also a time of hope. The need for speedy but reliable tests for hazardous chemicals is especially great, now that the EPA has the awesome task of assuring the safety of all new chemicals. Until 1977, only some drugs and the chemicals added to foods had to be tested. Then the Toxic Substances Control Act took effect.

Before passage of TSCA, (which is usually spoken as "TOSCA") chemicals were often found to be dangerous only after they had been manufactured and spread through the environment and after large industries had been built around them. This law has the goal of putting chemicals, not people, to the test. It is based in a principle recommended by a panel of the National Academy of Sciences, that "the burden of proof that society will obtain a benefit from a new use of a chemical should rest with those promising such use."

At best though, TSCA is only a halting step toward that goal. Intense lobbying by the chemical industry blocked passage of a strong law for five years. The fact that TSCA was eventually supported by the Chemical Manufacturers Association is testimony to its weakness. Manufacturers *are* required to prove the safety of new chemicals, but usually only after the EPA challenges the use of the substance and goes to court.

Dr. Sidney Wolfe, head of the Public Citizen Health Research Group, called the law "a very weak gate against the flood" of chemicals that has been pouring into the environment. He pointed out that TSCA gives the director of the EPA broad discretion in deciding which chemicals will be challenged and tested. "When there is discretion it inevitably flows to the power base—in this case, industry."

More than three years after passage of the Toxic Substances Control Act, the EPA had removed from the market just six new chemicals. TSCA had had little effect. Environmental and other public interest groups criticized the EPA for being too timid. An EPA official, weary of trying to persuade the chemical industry to cooperate, said that the law would probably have to be strengthened by Congress in order to do much good.

The need for testing chemicals—both old and new—is urgent, and the task is enormous. Further development of quick, reliable tests will certainly help. There is reason to be hopeful about improvement in both the quantity and quality of scientific information about health hazards. *Acting* upon the information involves business decisions and political decisions, and the choices may affect human health far into the future.

CHAPTER FOUR

How Safe Are the Water and Air?

An absolute prerequisite for good health is a dependable supply of reasonably pure water. Access to safe water is vital, and it cannot be counted on in many countries. As a result, such waterborne infectious diseases as typhoid, cholera, and dysentery are still major killers in the world.

In the United States we have taken for granted that our drinking water is safe. With rare exceptions it is free of such bacterial infections as typhoid. However, our society has been producing new chemical health hazards from such sources as automobiles, agricultural runoff, and industrial wastes, so we should not be surprised to find these chemicals in the water that is piped into our homes, schools, and businesses. A thorough analysis of the drinking water in Cincinnati, Ohio, detected 700 different chemical compounds, most of which had not been on the market twenty years ago.

Most of these compounds were probably harmless, or present in small, harmless amounts. In most communities,

There may be other hazards in water besides infectious diseases.

however, the water is not usually tested for such known hazards as viruses and pesticides, yet these contaminants are present in almost every sample when testing is done. The 40,000 water systems in the United States are not even equipped to test the water for dangerous chemicals, let alone keep them from reaching consumers.

Lead is the most common metal found in drinking water. It comes mostly from automobile exhaust gases and is washed into water supplies from the air, soils, and the leaves of plants. Zinc, iron, cadmium, and titanium are other metals often found in drinking water.

Nitrates are also present in many water supplies. Their main source is fertilizer, washed in from suburban lawns and agricultural fields. Excess nitrates, which cause a reduction of oxygen in the blood, pose a special threat to infants, who may die from the "blue baby syndrome." Several deaths in Minnesota were traced to nitrate-laden well water. Considering the growing national use of chemical fertilizers, more troubles with nitrates can be expected. There are ways to remove nitrates from well water or from entire water systems, but the methods are expensive.

Industrial pollutants appear in drinking water all across the nation. Among the compounds often found are trichlorethylene (TCE), a degreaser that causes liver cancer in mice. Tetrachlorethylene, a related compound, comes from the wastes of dry-cleaning plants. Low levels of benzene, vinyl chloride, and carbon tetrachloride, all known carcinogens, show up frequently in water tests.

In twenty-four California counties, the pesticide DBCP was detected in well water two years after the poison was banned in that state. Apparently, it took at least that long for the pesticide to travel from the land surface to underground water supplies. Every locality or region has its special brew of chemical hazards. In states having plentiful snowfall, excess amouns of salt have been found in water supplies. The salt comes from the 9 million tons that are annually spread on roads to help melt ice and snow. According to the EPA, in Snow Belt states about a quarter of the population drinks water contaminated with road salt, adding to an already oversalted diet for most people.

Application of fertilizers and pesticides to cropland sometimes contaminates drinking-water supplies.

Pollutants like salt, DBCP, nitrates, and lead cannot usually be traced back to an easily-identifiable place, such as the outlet pipe of a factory or a sewage treatment plant. These chemicals and many others are especially troublesome because they come from "nonpoint sources." They are washed by rainwater from streets, highways, suburban lawns, cropland, landfills, and construction sites.

All across the nation, especially in industrial areas, dumping places for toxic chemicals may also pose a threat to water supplies. Ninety-two billion pounds of chemical wastes are produced annually, but only a small fraction of this amount is disposed of properly or recycled.

The Love Canal in Niagara Falls, New York, drew nationwide attention in 1979 and 1980 when evidence was revealed that toxic wastes buried there were harming local residents. About 22,000 tons of chemical wastes had been buried in the abandoned canal and covered over with clay in 1953. In the late 1950s, a school and about a hundred homes were built on or beside the site. During the late 1970s, chemicals began oozing to the surface and seeping into basements. People noticed foul odors; some had trouble breathing. Since then, some of the more serious effects of the chemical wastes have become increasingly apparent: unusual numbers of women who suffer miscarriages, of children with birth defects, of adults with liver and genetic damage. Eventually about a thousand families were relocated from the Love Canal area at government expense. In this tragic case, people in the neighborhood were harmed by direct contact with toxic chemicals which were carried into their

Wastes may become health hazards long after burial.

CATERPILLAR
CAT

homes by ground water. Some of the wastes seeped underground to the Niagara River, but no harm has yet been definitely traced to them.

The Love Canal is but one example. The EPA estimates that health hazards may exist at anywhere from 1,200 to 34,000 dumping sites, storage sites, or landfills that contain hazardous wastes. Seventy-five percent of all landfills are located in areas where chemicals can travel through the soil and reach rivers, lakes, and wells. Properly designed landfills are supposed to prevent this spread, but only time will tell whether they will; some of the chemicals last a long time and may eventually escape, decades from now.

Soil samples are collected (left) for chemical analysis at Love Canal in Niagara Falls, New York. Many families were evacuated to protect their health.

Almost every state has some trouble spots. A landfill on Long Island, New York, caused pollution of a billion gallons of groundwater and thc closing of many wells. Improperly stored PCBs (for details about PCBs, see Chapter 6) were found just 150 feet from wells that supply water to Youngsville, Pennsylvania. And in Iowa, traces of arsenic and TCE from a chemical dump reached a river that is the source of drinking water for two cities downstream.

So far, about 4,000 cases of sickness a year (mostly infectious diseases) can be directly linked to drinking water. The untraceable harm may be much greater. A study by the Environmental Defense Fund showed a possible link between unusually high cancer rates in certain parts of Louisiana and water supply. The Louisiana communities took their water from the Mississippi River. Even after being treated for use, the water still contained chloroform, benzene, and other carcinogens.

In 1977, the EPA set maximum allowable levels in water for such chemicals as lead, arsenic, and toxaphene. But no limits were set for scores of other chemical compounds found in water supplies, party because of lack of information about their effects on people, but mostly for political reasons. Learning that a chemical in the environment is harmful to humans is an easy step compared with the task of controlling it or restricting its use. The EPA and other regulatory agencies are so undersupported by Congress that bold action is discouraged. The first step toward restricting a toxic substance usually leads to resistance from industry, followed soon after by resistance from Congress. All too often, there is the possibility that Congress will weaken the law or find another way for industry to sidestep it. (Thus,

Methods of treating tap water have not changed for decades, although new kinds of contaminants have appeared.

after attempting to ban saccharin, the FDA found its basic food-safety law under congressional scrutiny.) The safe approach for regulatory agencies is to do little or nothing. This inertia can be overcome somewhat by strong support from a president, but the main pressure for change comes from individual citizens and from consumer and environmental groups, acting in the public interest.

Persistent pressure from the Environmental Defense Fund caused the EPA in 1978 to propose new regulations to remove organic chemicals from drinking water. (In general, organic chemicals are those involving carbon compounds. They include such chemical groups as alcohols, ethers, esters, and aldehydes and some chlorides and amines.) A specific limit (100 parts per billion) was proposed for chloroform and other organic chemicals of a group called triholomethanes. These chemicals form when chlorine is added to water supplies and reacts with dead leaves and other organic matter. Chlorination is, of course, *the* traditional way to disinfect water and is used in almost every water system in the United States.

The EPA's move did not condemn the use of chlorine but called for more exact control of the process and possible use of other antibacterial agents and methods—as long as the triholomethanes were reduced. These unintended by-products of chlorination are carcinogens.

The EPA also sought to reduce potentially dangerous chemicals in water by proposing filtration through granulated carbon. This method had been used successfully in Europe. Most organic chemicals are adsorbed on the carbon as water flows through. Unpleasant odors and flavors are also removed.

These proposals represented the first major change in water treatment in more than fifty years. As you might expect, resistance came from the waterworks industry. Steven Marcus, an engineer writing about this industry in *Technology Review*, observed that "its first reaction to the proposed regulations—in the grand tradition of polluters and negligent pollution controllers—was to deny the problem outright, to impugn the motives of those who take it seriously, and to call for more study."

Some unexpected resistance came from the Council on Wage and Price Stability, an administration anti-inflation group, which charged that the cost of change in water systems, estimated at $700 million, was not justifiable. To this sort of thinking Dr. Robert Harris, a water quality engineer with the Environmental Defense Fund, replied, "We are sitting on a time bomb. The implications of not acting now to clean up the water are far greater than the implications of acting and spending a relatively minor amount of money. If in twenty years we find that the epidemiological studies were wrong, all we will end up with is better-tasting and smelling water. If we are right about the danger and we don't act, the health consequences could be disastrous."

If the EPA proposals are to be implemented, lawsuits and other delaying tactics by the waterworks industry must be overcome. Meanwhile, the agency and consumer groups warned people not to rely on home water filters, which can become breeding places for bacteria and can be more dangerous than no filter at all. Another common stopgap, bottled water, is expensive and sometimes contaminated with bacteria. In the long run, organized political action is the most effective way for citizens to get a safer water supply for their community.

TROUBLE IN THE AIR

We can avoid hazards from foods or drugs by not using those that are known to be harmful. But avoiding health hazards in the water supply is difficult because there is just one supply. The same is true for the air—just one supply—and we have to keep breathing. Airborne hazards are similar to other kinds, though. The concern and information about most of the harmful airborne substances are relatively new, the scientific understanding is incomplete, and there are often social, economic, and political obstacles to overcome in order to reduce the risk to people.

The most hazardous pollutant people can breathe is

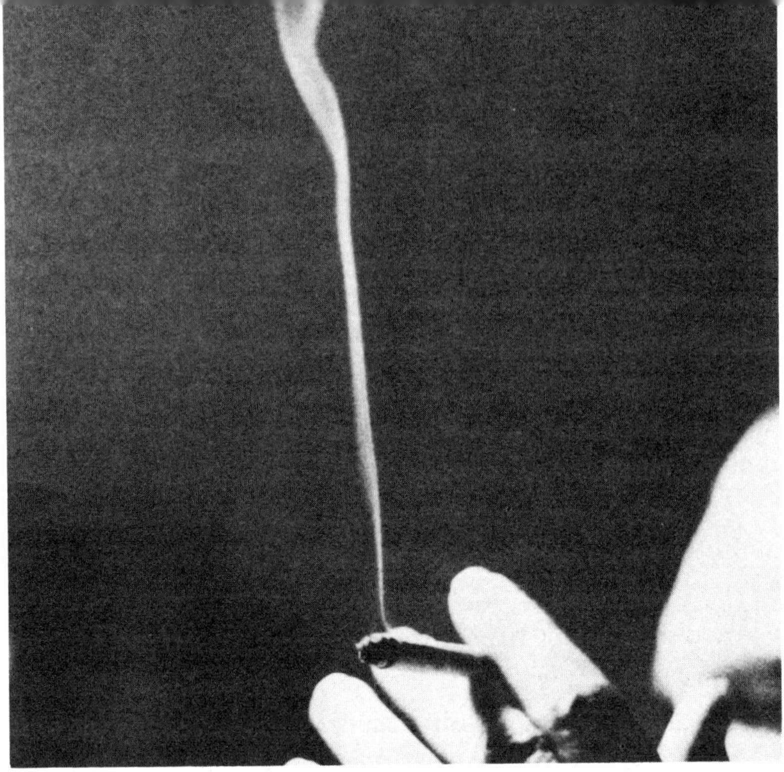

An idling cigarette is dangerous to both smokers and nonsmokers.

usually, but not always, taken voluntarily via smoking. Tobacco smoke contains not only carcinogens but also substances which promote cancer formation. Cigarette smokers are at least ten times as likely to develop lung cancer as nonsmokers. The risk of lung cancer increases enormously when the smoking habit is combined with another hazard, such as taking birth control pills or working in the asbestos, rubber, uranium, or chemical industries. And of course the health troubles aren't limited to cancer. A middle-aged male smoker is twice as likely as a nonsmoker to suffer a heart attack.

Widespread knowledge about the dangers of smoking has had some effect. Many doctors have stopped smoking, and the proportion of adult smokers has fallen markedly in the United States, Britain, and several other nations. In these countries and in developing nations, however, there is a troubling trend: an increasing percentage of teen-age girls and young women who smoke.

Women of childbearing age risk not just their own health but that of their babies. The infant mortality rate is greater for babies of smoking mothers than for nonsmoking mothers. And the infants of women who smoke weigh less, on the average, than infants of nonsmokers. A 1979 report issued by U.S. Surgeon General Julius B. Richmond concluded, "Cigarette smoking, an early sign of woman's social emancipation, is now a major threat to her personal health and her ability to bear children."

In the late 1970s, controversy began to grow over involuntary or passive smoking—the inhalation of cigarette smoke by nonsmokers. Past complaints of nonsmokers had had little effect; to smokers, they seemed like troublemakers, interfering in the rights of others. Then evidence began to accumulate that the health of nonsmokers can be affected by cigarette smoke. Two-thirds of the smoke from a burning cigarette enters the general environment. A cigarette may burn for ten minutes, but a smoker inhales from it for no more than a half-minute. Analysis of smoke from an idling cigarette shows that it is more dangerous than that which a smoker inhales directly.

This sidestream smoke, as it is called, contains more cadmium, twice the amount of tar and nicotine, three times more of the carcinogen benzopyrene, five times more carbon monoxide, and fifty times more ammonia than smoke that is inhaled by a smoker. So nonsmokers are exposed to these harmful substances in the air, plus leftovers exhaled by smokers.

In some professions, bartending, for example, nonsmoking workers are sometimes forced to inhale smoke which is the equivalent of smoking a pack of cigarettes a day. But you don't have to be a bartender or waitress in order to suffer ill effects from secondhand smoke—just spend thirty minutes in a smoky room, or live with a smoker. According to studies in Britain and Israel, parents' smoking doubled the risk of bronchitis or pneumonia in their infants. There is some evidence that children of smoking women also have deficiencies in physical growth and mental development. People who live in the home of a smoker also suffer from more colds and other respiratory infections than normal.

Because of such evidence, progress has been made toward reducing involuntary smoking in public. Nonsmoking areas have been established in airplanes, trains, restaurants, and many other public places. Thirty-two states have passed some sort of law aimed at providing smoke-free zones in enclosed public places. A tough smoke-limiting law on the 1978 ballot in California was defeated, however, after the tobacco industry spent over $5 million in its advertising campaign in that state. The tobacco industry has been called "dollar for dollar, probably the most effective lobby on Capitol Hill." For nonsmokers, the right to air free of tobacco smoke will not come easily.

One ingredient of cigarette smoke, cadmium, is a source of increasing concern among health experts. This metal is used in electroplating, solder, vapor lamps, nickel-cadmium batteries, incandescent light bulbs, and the manufacture of plastics. From these products and their wastes cadmium reaches the air, water, and soils. Cadmium is also produced during the burning of oil and coal. It is taken up by some crops, including tobacco plants. Smokers have cadmium levels in their bodies that are about double those of nonsmokers.

As is often the case, the first evidence of cadmium's harmfulness showed up among factory workers who were exposed to high doses of the metal. Inside the body, most cadmium accumulates in the kidneys and liver. It causes kidney disease, lung damage, and perhaps hypertension. Symptoms of cadmium poisoning include chronic bronchitis and emphysema. Long-term exposure to cadmium may lead to what the Japanese call *itai-itai* (ouch-ouch) disease, in

Compared with a normal lung (left), the lung of a heavy smoker (right) has greatly enlarged air sacs.

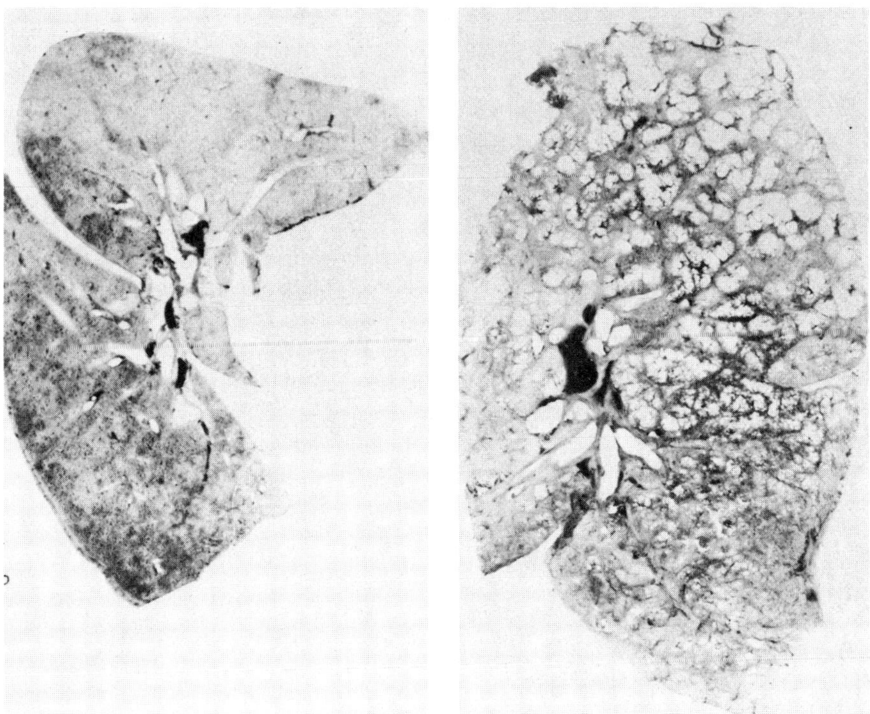

which damage to the kidneys prevents the normal supply of vitamin D from reaching the bones. They become brittle and break, causing a slow, painful death.

Cadmium is excreted very slowly. If it reaches toxic levels in the body, there's little that can be done about it. According to Dr. David Rall, director of the National Institute of Environmental Health Sciences, "The cadmium levels in the kidneys of the general population are a third to a half those levels that we know are toxic. If, as we suspect, cadmium levels will increase, we're particularly concerned about the future."

Research in Sweden revealed high levels of cadmium in sewage sludge and in ash and slag from incinerators. Ninety percent of the cadmium came from plastic and metal products in household trash. Alarmed about the amount of cadmium in the environment, and in its citizens, the Swedish government banned most products containing cadmium in 1980.

Lead is another metal that accumulates in the body, and which can reach us through the air we breathe, as well as via water and food. It was one of the first metals mined and used by humans and has been known to be poisonous since the time of the ancient Greeks. As knowledge of the dangers of lead has grown, such uses as lead-based paints, lead water pipes, and lead glazes for pottery have been reduced or abandoned.

Today, lead is used in electric-storage batteries and as an additive to gasoline. Nearly all of the lead in the air comes from gasoline. Studies of the lead deposits in glacial snows

Leaded gasoline is the major source of lead pollution.

in Greenland reveal that heavy lead contamination began after 1940, coinciding with increased use of leaded gasoline. Lead pollution is global and in all foods, especially in foods stored in lead-soldered cans.

High doses of lead cause irreversible brain damage. Lower doses cause malfunctioning of many body systems. But there is uncertainty about the overall harm of low doses and whether any level of lead in the blood can be considered safe.

In 1978, after a court order was secured by the Natural Resources Defense Council, the EPA finally set a national legal standard for lead: 1.5 micrograms of lead per cubic meter of air. Cities or states with higher levels had to take steps to reduce it, perhaps by restricting the use of automobiles, or the amount of lead added to gasoline. But environmental and consumer groups claimed that this level was too high, especially for children, the people most vulnerable to harm. A 1980 report by the National Academy of Sciences stated that "growing evidence suggests that commonplace exposures to lead in urban environments may be associated with detrimental effects on children's intellectual development and behavior."

In a study involving more than 3,000 elementary school children near Boston, scientists found that amounts of lead well below toxic levels interfered with learning. (The children were not given lead by the scientists but had acquired it from their own environments.) Other studies have linked lead to hyperactivity and learning disabilities in children. Some scientists believe that even minute amounts of lead can affect the behavior and learning of children and that stricter limits must be placed on the sources of lead now entering the environment.

Asbestos is another airborne health hazard which may threaten children and students because it was once widely used in school buildings. The name *asbestos* refers to a group of crushed silicate minerals made up of tiny fibers. For many years, this practically indestructible material was called the "miracle mineral" because of its resistance to heat and fire. Community building codes insisted that asbestos be used in schools, offices, and many other structures. Between

Many children in urban environments may be adversely affected by lead in the air they breathe.

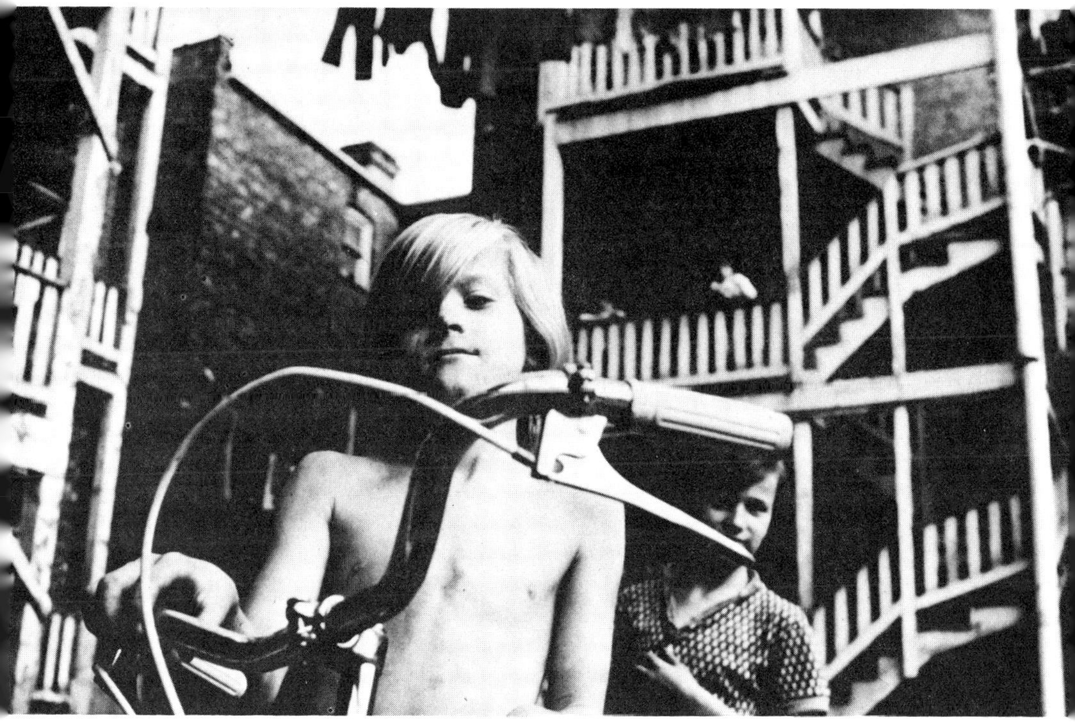

1940 and 1973, it was sprayed on the walls and ceilings of countless buildings as fireproofing, soundproofing, or mere decoration.

The bad news about asbestos began to unfold in the early 1950s when Dr. Irving Selikoff, now of the Mount Sinai School of Medicine in New York City, opened a clinic in New Jersey. The clinic happened to be near an asbestos products plant, and some of its workers were his patients. Through the years, Dr. Selikoff noticed an unusual incidence of lung cancer and also of mesothelioma (an always-fatal cancer of the lining of the chest cavity or abdomen) among the workers.

By the early 1960s, Dr. Selikoff had enough evidence, and concern, to begin a major epidemiological study of asbestos workers. He and his colleagues were at first thwarted by uncooperative asbestos manufacturers. Help came from labor unions, however, and Dr. Selikoff was able to track down thousands of present and former asbestos workers. In many cases, the investigation lead to death certificates. Unusual numbers of the workers were dead, and the rate of lung cancer and mesothelioma was exceptionally high, not just among workers but also among members of their families, who inhaled asbestos fibers from clothing worn home from the factory.

Only in 1978 did the federal government issue a public warning about asbestos. Efforts have been made to protect workers, but the general public is still exposed to varying levels of asbestos. Sources include asbestos dust that is released from auto brake linings, fibers that enter the air when buildings are torn down, and fibers that flake off ceilings and walls of existing buildings.

State health officials throughout the country searched for asbestos in schools and other buildings. The eventual cost of removing asbestos, or covering it with an airtight sealant, will be enormous. It seems likely that there will be other, unknown costs and public alarms, since asbestos is used in many products. The use of toasters and iron-board covers, which often contain asbestos, probably present no health risk to people. But what about a product that blows asbestos fibers into a person's face during use?

In 1978, investigative reporters for television station WRC-TV in Washington, D.C. discovered that hand-held hair dryers were commonly made with asbestos heat shields. The station informed the Consumer Product Safety Commission, but that agency showed no interest, mistakenly believing that asbestos was no longer used in hair dryers. The television station paid for a study which showed that asbestos fibers sprayed out of hair dryers. Broadcast of the results caused a national furor and led to the repair or replacement of millions of hair dryers. This incident also drew attention to the poor record of the Consumer Product Safety Commission which, since its creation in 1973, according to one reporter, "has a reputation for bureaucratic torpidity that is unmatched in Washington."

OTHER DANGERS IN THE AIR

Just how dangerous is air pollution? One study found a much higher than normal rate of lung cancer among men living in areas where the petroleum industry is concentrated. Certain hydrocarbons given off during oil refining seemed to be the likely cause. Overall though, city dwellers have about

the same chance of getting lung cancer as country dwellers, unless they are cigarette smokers. Smokers are more likely to get cancer in an urban area because the pollutants seem to act synergistically with the carcinogens in tobacco smoke.

The most troublesome pollutants in the air include sulfates, which irritate the human respiratory tract. Low concentrations aggravate such diseases as asthma, bronchitis, and emphysema. The main source of sulfates is sulfur dioxide, produced by burning such sulfur-bearing fuels as coal and oil. Once in the atmosphere sulfur dioxide goes through a series of changes to form sulfates. Vehicles that burn diesel fuel are particularly troublesome, since the wastes from burned fuel contain much more sulfur dioxide and tiny soot particles than a gasoline-powered vehicle. Most of these soot particles are small enough to be breathed deep into lungs. The soot contains chemicals that are known to cause cancer and to increase the risk of bronchitis, asthma, pneumonia, and emphysema.

Nitrates also occur in the air as small particles and have been linked with increased asthma attacks and other respiratory illness. Nitrates come from nitrogen oxide compounds, another waste from burning fossil fuels.

Some kinds of pollutants, especially the visible "smokestack" kinds, have been reduced. Overall, however, progress toward reducing air pollution has been slow, mostly because of the political power of auto manufacturers, power companies, and such industries as steel, copper, and lead.

Air-sampling balloons are released in southern California.

Strong support from the White House and from the public is needed to push the EPA toward vigorous pursuit of healthier air. The EPA is under increased pressure from environmental and health groups to develop regulations that limit the amount of fine particles and other pollutants in the air we breathe. With greater amounts of coal being burned, more effort must be made just to maintain present air quality, let alone improve it.

Our worries about air pollution usually stop at the door of our homes, as though the air inside was somehow protected from harmful substances outside. It isn't. A 1978 study in the San Francisco area found that the air inside the average home was often more polluted than the outside air on a smoggy day. The indoor levels of such pollutants as carbon monoxide, nitrogen oxide, vinyl chloride, formaldehyde, and fluorocarbons were often two to three times higher than the air quality standards set for outside air by state and federal governments. The pollutants came from aerosol sprays, cigarette smoke, foam insulation, compounds used in cleaning and cooking, and from gas appliances, especially stoves. In one experiment, use of a scented oven cleaner left detectable amounts of thirteen organic chemicals in the kitchen air. Another dangerous substance is radioactive radon, which is present in varying amounts in brick, concrete, stone, and tap water. Homes tested in Maine, Montana, and Florida had indoor concentrations of radon several times greater than those found normally outdoors.

This discovery is troubling enough, but changes in home construction may make the problem worse. In most well-built single-family homes, the air is completely replaced by air from outside in an hour's time. In order to conserve energy, however, people have been adding insulation and taking other steps to make their homes more airtight. A national building code for all new homes is being developed, too, with the goal of reducing the air exchange rate to one change in two hours or even one change in four hours. Achieving this goal will certainly save energy, and heating and cooling costs, but may also allow dangerously high levels of pollutants to build up indoors. Since the kitchen is usually the most polluted room, the new building code may have to include a kitchen fan which vents air outdoors. It is especially important to vent gas-burning stoves outside. Materials that emit radon should be painted with sealants. And homeowners may want to install heat exchangers (which are common in Scandinavian countries). They pull fresh air in and blow used air out, but help conserve energy by allowing heat from the used air to warm incoming air.

Home air pollutants are usually concentrated in the kitchen.

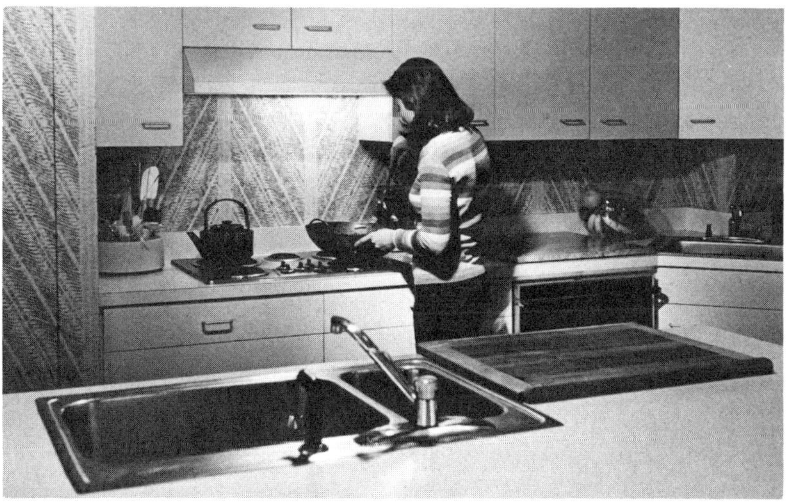

Human health and property values are adversely affected by noise along airport flight paths. Many rock musicians (right) and discotheque disc jockeys suffer serious hearing loss.

Not all health hazards in the air are inhaled. Noise, for example, reaches our ears as sound waves. Compared with other environmental health hazards, noise pollution has received little notice, yet it takes quite a toll. The EPA estimates that 16 million people in the United States have already suffered hearing loss caused by noise, and another 40 million (not counting factory workers) are exposed to hazardous noise levels.

Partial loss of hearing is just one effect of high noise levels. Loud noise may be even more damaging to the nervous system that to hearing. A study of the medical records of factory workers showed that individuals who worked in noisy areas had more sickness and accidents than those who worked in quieter areas. Noise makes people tired, nervous, irritable, and antisocial. The noise causes stress which in turn causes a rise in heart rate and blood pressure and has other effects on body systems. Noise-related stress is suspected of contributing to heart disease, diabetes, and arthritis.

Most of the physical and emotional effects of noise can be achieved by the sound levels made by common home appliances. Blenders, vacuum cleaners, electric shavers, and hair dryers all produce noise in a range considered potentially harmful to people. The EPA has the authority to set noise

limits on appliances and other machines but has not done so. Manufacturers claim, of course, that any helpful change would be too expensive, although some already make quiet models to meet requirements in Europe and sell noisier ones in the United States. One obstacle to a quieter environment is that people expect some machines to be noisy; a quiet model of a vacuum cleaner failed to sell well because people equated low noise with low cleaning power.

LOW-LEVEL RADIATION

Another possible health hazard in some homes is radiation from microwave appliances, although the scientific evidence is particularly inconclusive on this matter. Two million microwave ovens are used in homes and businesses. Microwave radiation is also used in such devices as burglar alarms, smoke detectors, and garage door openers. All of these devices do not normally emit the high levels of microwaves that are capable of harming people, but some scientists claim that the long-term low doses of microwaves can reduce immunity to diseases and also cause infertility, birth defects, behavioral problems, and cataracts (clouding of the lens of the eye). Other scientists say that the level of microwave radiation to which people are exposed is far too low to cause any injury.

In a 1978 report, the General Accounting Office (GAO) concluded that federal agencies had been lax in protecting the public from potential harm of microwave radiation. In its review of over a hundred research reports, the GAO found marked disagreement about the safety of the present microwave standard, set in the 1950s. This standard is not even a legal government one, and the GAO urged that mandatory exposure standards be set to protect the public and workers. Before such standards are set, however, it seems that the limited research on microwave radiation will have to be augmented, especially to learn more about long-term, low-level effects on living things.

There is another kind of radiation that now worries people much more, partly because it is more widespread and partly because there is ample and growing evidence of its harmfulness. This is ionizing radiation, which includes x-rays and all types of radiation given off by radioactive substances. All ionizing radiation has the potential of damaging cells and eventually causing leukemia and other cancers, as well as birth defects in later generations.

People have always lived with some low levels of radiation. There is radioactivity in the food we eat, the air we breathe, and even in our bones, from radioactive potassium. Radiation also comes from space in the form of cosmic rays and from the earth's rocks and soils in the form of gamma rays. This natural or background radiation varies from place to place; the world average is 100 millirems a year. (A millirem is a thousandth of a rem; a rem is a unit for exposure to radiation which stands for "roentgen equivalent, man.") Medical and dental x-rays are the major source of human-made radiation. And even though few nations now test atomic weapons aboveground, some radiation remains inside people who were exposed to fallout from tests conducted in the 1940s and 1950s. Low levels of radiation also are released into the biosphere during every stage of the fuel cycle of nuclear power plants, from mining to waste storage.

New evidence suggests that low levels of radiation are

more dangerous than previously thought. In 1979, for example, Dr. Irwin Bross, chief of biostatistics at Roswell Park Memorial Institute, a cancer research center in Buffalo, New York, reported that people exposed to low levels of radiation at work or from frequent x-rays may be ten times as likely to get leukemia as had been thought. And in 1979, Scottish scientists also reported the discovery of radiation-caused genetic changes in the white blood cells of men who refueled nuclear submarines. The men had been exposed to doses *below* the limit (5 rems a year) believed to be safe for their occupation.

More evidence came from an unplanned long-range study—the exposure of more than a half-million soldiers and civilians to radioactive fallout from 183 nuclear weapons tested between 1946 and 1964. Troops often went on war maneuvers near the test sites. They were assured that there was no risk. Residents downwind from the blast sites were told the same thing.

This was, at best, wishful thinking. Although the United States Army kept poor records and investigators have had difficulty locating many former soldiers, there is already proof of unusual incidence of leukemia among the troops who were near the tests. There is also no doubt that leukemia among Utah children increased, with the highest rates coinciding with areas and times of greatest fallout. Data on this were available in 1965 but were suppressed so as not to interfere with government promotion of nuclear energy.

During 1979 congressional hearings on the bomb test

Soldiers watch a 1951 nuclear bomb test in Nevada.

aftereffects, a Nevada woman whose son had died of leukemia said, "We were feeding our children and families poison from these bombs. We were forgotten guinea pigs. At least *real* guinea pigs are checked."

In earlier testimony, Dr. Karl Morgan, a professor at the School of Nuclear Engineering of the Georgia Institute of Technology and an expert on the health effects of radiation, said that there is no safe level of radiation. He said that he was appalled at the scientists who, in spite of "an overwhelming amount of data," continue to claim that present radiation standards are safe. He identified these scientists as mostly being associated with the United States Department of Energy.

Revelations like this have added to a wave of concern and protest about radiation. In some states, mostly in the West, people worried about radiation from uranium ore tailings (mine wastes) which were left in piles and sometimes used in construction of homes and other buildings. The nation's commercial nuclear energy program came to a virtual standstill, partly because of high costs, plant safety problems, and a slowing growth of electricity use, but also because of the unresolved matter of safe disposal of highly radioactive and long-lived wastes.

Many millions of dollars had been spent on studies of nuclear waste storage, but the focus of the research was quite narrow. It was based on the assumption that the wastes would be solidified, sealed in glass, and buried deep in salt beds. For decades, the public and Congress had been assured that this method was the safest and would be carried out. Eventually, however, independent geologists and other scientists were encouraged to speak up, and they found many flaws in the old assumptions. One geologist called the glass-salt storage idea "the worst possible combination." Research was broadened to consider storage in ceramic materials and in other kinds of stable geologic deposits.

Senator Gary Hart, chairman of the Senate subcommittee on nuclear regulation, said, "If the word 'scandal' can be attached to nuclear power, it is that this industry has been permitted to expand for two and a half decades without an acceptable solution for waste disposal."

For the time being, at least, nuclear wastes are mostly a potential hazard and not a significant source of radiation in people's daily lives. The greatest single source continues to be x-rays. Their use has become a basic part of the practice of medicine and dentistry. In 1978 there were 161 million x-rays taken for medical purposes and 86 million for dental purposes in the United States. Critics say that many of these are unnecessary. Furthermore, a 1979 GAO study showed that more than a third of dentist's x-ray units gave "excessively high" exposures. It is ironic that x-rays, which can help detect and combat cancer, can also cause it.

Radioactive materials are also used to treat cancer and other diseases, especially to kill cancerous cells. At least 700,000 people each year receive radioactive iodine for thyroid disorders. This nuclear medicine is vital, but doctors often give more radiation than necessary. Some patients give off such high amounts of radiation that they increase the risk of thyroid cancer for members of their family.

In 1979, Joseph Califano, then Secretary of Health, Education, and Welfare, directed the FDA to increase its efforts to reduce human exposure to x-rays and nuclear medicine methods. He also urged people to be wary of exposing them-

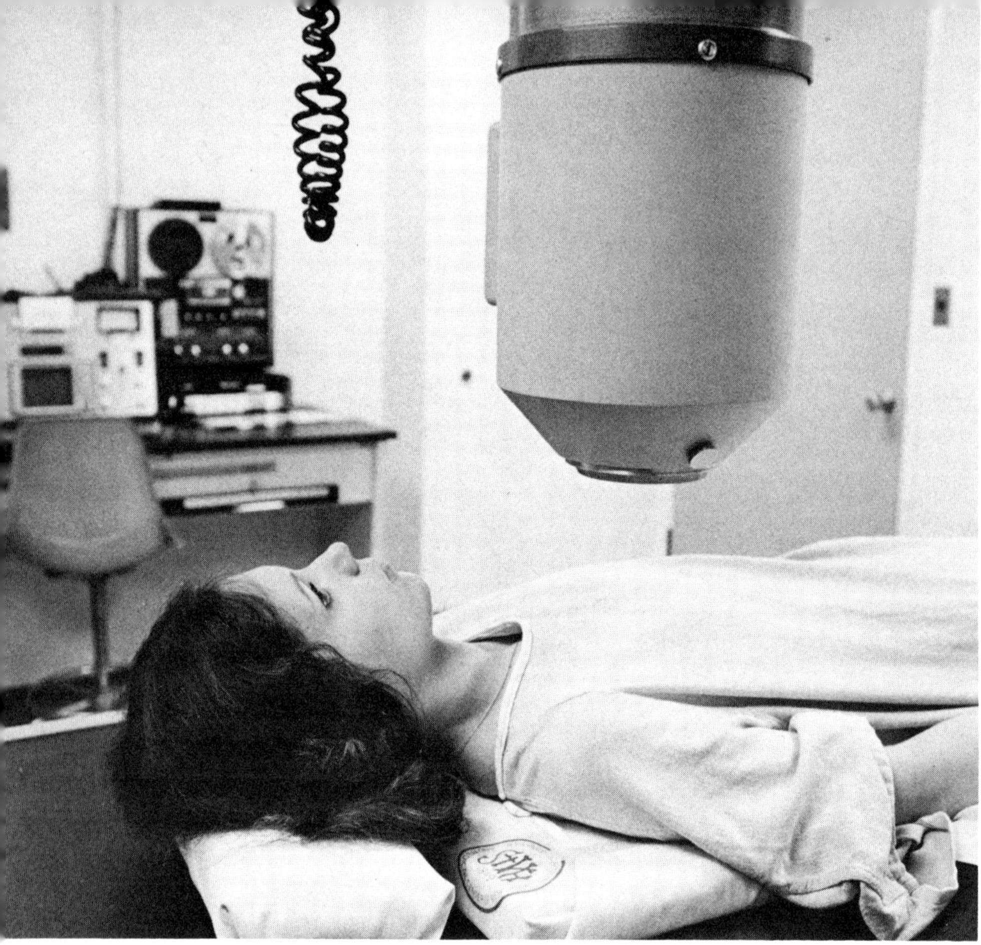

The ability to take x-rays from several angles makes the "CAT scanner" an invaluable device for detecting illness.

selves to low levels of radiation. The existing legal levels for radiation exposure would not be lowered, Califano said, because more research was needed. To reduce the permissible levels would imply a quality of understanding that did not exist, Califano believed. Some scientists agreed. Others believed that there was already sufficient evidence to reduce radiation limits. Nearly all agreed that the government's overall program concerning low-level radiation remained disorganized and ineffectual.

Although radiation sometimes reaches people by rather roundabout routes, other human-made substances drift beyond the atmosphere to cause a possible threat to health on earth. These chemicals are chlorofluoromethanes (CFMs), also known as chlorofluorocarbons or by the trade name Freon. Use of CFMs to spray products from aerosol cans was banned in 1979, because the gases rise into the stratosphere and cause a reduction in the ozone that surrounds the earth. The less ozone, the more ultraviolet radiation reaches the earth's surface. Increasing ultraviolet radiation may have several indirect harmful effects on people, for example, changing the climate or causing a lower yield of food from crops and ocean life. It can also affect people directly, causing increased skin cancer.

The most common form of skin cancer is rarely fatal, but another kind—melanoma—is often fatal and its incidence has been growing. Cases of both types will increase sharply, according to a 1979 report by the National Academy of Sciences, unless the loss of ozone is halted.

The cause of the trouble—CFMs—no longer escapes from aerosol cans in the United States, but does elsewhere in the world. Furthermore, CFMs have many uses, in commercial and home refrigeration and air-conditioning systems, in auto air conditioners, and as blowing agents for plastic foam products such as those used in seat cushions. Substitutes must be found and used, and not just in the United States, for this is clearly a global problem. Further delay, according to the National Academy of Sciences, will only increase the eventual problems. The same can be said about many other substances in the air around us and in the water we drink.

CHAPTER FIVE

America's Deadly Diet

In the good old days, only a few years ago, bacon and eggs were thought to be a hearty and healthy breakfast. For dinner, a juicy steak was another sign of eating well in the United States.

Today we know that eggs are high in cholesterol, a substance that may be linked with heart disease, most bacon contains a food preservative that may lead to cancer, and steak from grain-fed beef is especially high in animal fats, another substance linked to heart disease.

We also know that the typical United States diet, so rich in animal protein, fats, salt, and sugar, is not good for humans. One-fifth of all Americans are so overweight that their health is threatened, especially by heart disease, the nation's number one killer. The great majority of diabetes is due, in part, to overnutrition. And some of the most common cancers have been linked to diets high in animal protein and fats. Rather than directly causing such cancers, substances in the diet somehow encourage their development—the typical American diet is cancer promoting.

The evidence of this disturbing news comes from a variety of sources, including studies of Seventh Day Adventists, who, for religious reasons, eat a low-fat, vegetarian diet. As a result, they have much lower rates of cancer of the colon, breast, and uterus than other people in the United States. Studies of the human digestive system suggest that people are not well suited to a heavy meat diet. Evidence from anthropology and archeology indicates that humans, through most of their four-million-year existence on earth, have been primarily plant eaters, not meat eaters.

Indeed, some scientists believe that a change in lifestyle, particularly in diet, is the key to reducing the toll of heart disease, cancer, diabetes, and other common killers in rich nations. "In creating its way of life, each society creates its way of death," wrote Erik Eckholm of the Worldwatch Institute.

Giving up smoking, eating fewer hamburgers and less sugar—these are important steps toward better health, but there are sound reasons for concern about *all* of the contents of our food, drinks, and drugs. Life is hazardous enough without deliberately adding more dangers to our daily diets and our medicine cabinets, especially when the substances provide trivial benefits, or none at all.

Since 1958, federal law has required that food additives be tested by manufacturers and that no hazardous chemical be added to food. One legal loophole established at that time was the GRAS list ("generally recognized as safe"). It is a large loophole; any additive that appears to be safe in common use can be added to the GRAS list without testing. Along with saccharin, some other chemicals placed on the GRAS list have since been proved to be harmful.

Many food additives are beneficial and some are vital—to protect food from spoiling, for example. Others are of questionable value or have harmful effects which may outweigh their benefits. The food industry boasts of the vitamins and minerals it adds, but an excess of these substances can do harm. Too much iron, for example, can cause a serious illness called hemochromatosis in men. The disease causes changes in skin color, and contributes to diabetes, liver troubles, and heart failure. In Sweden, where 42 percent of the iron in the diet comes from food additives, 2 percent of the men examined in a screening program were found to be in early stages of hemochromatosis. In the United States, about a quarter of the iron that people consume is from iron-fortified foods. Many people are getting more iron than they need, and doctors have been alerted to watch for signs of this serious form of overnutrition.

There are already many volumes of evidence about the

Most of these foods contain sugar; all contain salt.

harm caused by excess sugar and salt. These additives plus corn sweeteners (corn syrup, dextrose, and fructose) make up 93 percent by weight of all food additives in the United States. Besides the products we usually think of as sugary—candy, for instance—crackers, salad dressings, peanut butter, and countless other foods have sugar added. The same is true of salt. Manufacturers say that they are only giving customers what they want, ignoring the fact that people have no real choice. People have come to expect the taste of excess sugar and salt because it is often difficult to find products without these additives.

Efforts have been made to protect children from heavily-sugared cereals, with little success. Some states and cities have taken steps to prevent students from eating so-called junk foods by banning sales of these foods in schools. In 1980, the United States Department of Agriculture (USDA) established a similar national program. More than 25 million children receive school lunches paid for in part by $3 billion of federal tax funds. The USDA program bars sales of junk foods in schools until after the last lunch period. It defines such food as that which has less than 5 percent of the recommended daily allowance of any of eight basic vitamins and other nutrients. This, said critics of the plan, was a loophole you could drive a candy truck through. A candy bar containing a single nut would meet the standards proposed by the USDA. And it would be cheap and easy to add enough vitamin C to change any junk food item into a "fortified food" that would qualify.

The USDA has taken steps to reduce fat, salt, and sugar in foods it helps schools buy for student lunches.

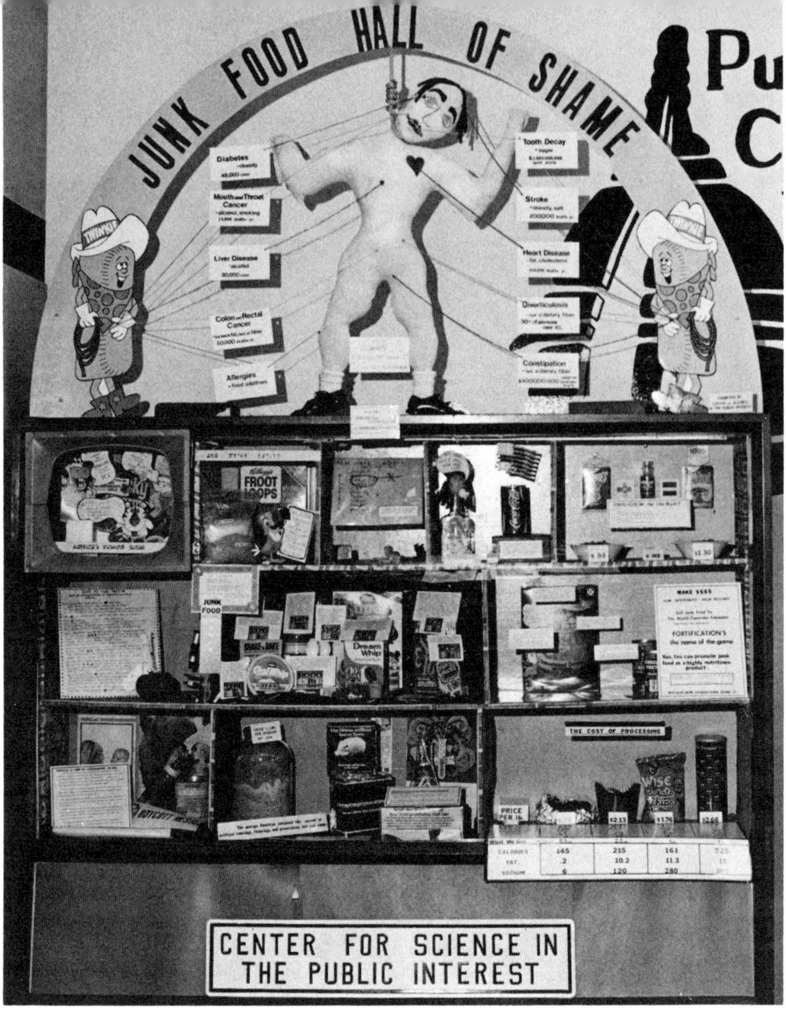

Criticism of some food industry practices was presented in a Junk Food Hall of Shame, prepared by the Center for Science in the Public Interest.

To consumer advocates, the USDA proposal seemed designed so as not to offend the snack food industry. Dr. Joan Gussow, director of nutritional education at Columbia University's Teacher's College in New York City, said, "If you are trying to sell a nutritionally sound lunch, you do not compete with yourself by offering other foods. After all, taxpayers are supporting a program to provide good food for their children. Then they have the program undercut with the sale of food that brings profit to many outside groups. It seems ridiculous. Fortification of these foods does not deal with the basic problem of having too high a consumption of salt, fat, and sugar."

Actually the new USDA policy included steps to deal with the latter problems. For the $1 billion in food the department buys directly for school lunches, it insisted on less fat in ground meat, less sugar in canned fruits, and less salt in canned meats than before.

Although sugar has long been linked with tooth decay, obesity, and diabetes, only recently have many citizens become aware of the dangers of salt (sodium chloride). Throughout nearly all of human history, people ate no salt other than that which was naturally present in foods. The human body evolved under conditions where sodium was scarce. Human kidneys are adapted to conserve sodium, so an excess accumulates in body fluids. This leads to swelling of tissues, and also to an increase in blood volume, blood pressure, and heart rate. Medical experts consider sodium from salt to be the major cause of high blood pressure (hypertension). Twenty-five million Americans have this disease, which can lead to kidney failure, stroke, and heart disease. There is little hypertension in nations where people have a low-salt diet.

Some people are more susceptible than others to serious harm from excess salt, but no one benefits from it. People need at most 5 grams of salt a day. Two grams are probably enough for nearly everyone. But from 10 to 24 grams of salt a day are consumed by most people in the United States. Some of this is consumed voluntarily, for example, when we sprinkle salt on food or add it to a recipe. But most of our salt is eaten involuntarily. It is hidden in foods. Actually it is not truly hidden, since manufacturers of many kinds of foods are required to list the ingredients on the packaging, in order of their quantity. Salt appears often and near the top of the ingredients' list—in ketchup, bread, cereals, puddings, pancakes, canned soups, cured meats, bouillon cubes, and nearly all canned vegetables.

Salt helps keep food from spoiling. It also masks unpleasant tastes and is a substitute for flavors lost in food processing. All of these problems can be solved without using salt, for example, by seasoning foods with spices, herbs, garlic, and onion. The tremendous salt excess in American diets is partly a result of people liking the taste or becoming used to it. Manufacturers are quite willing to continue oversalting foods because it seems to be good for business, if not for human health. Only an extraordinary effort by consumer

Season foods with these substances, rather than salt.

groups caused baby-food makers to reduce the amount of salt in their products. An even greater effort will be needed to bring about a reduction or elimination of salt in other processed foods.

In addition to salt, sugar, and corn sweeteners, the average American eats more than nine pounds of other additives a year, in the form of about 4,000 different chemicals. The additives include preservatives, colors, flavors, vitamins, thickeners, anticaking agents, curing agents, and emulsifiers (which help substances mix and stay mixed). Most of these are synthetic chemicals, but a few are natural. Carotene, for example, is the source of vitamin A in carrots and is sometimes used to color dairy products.

Food dyes are additives of questionable value since they affect only the appearance of food. They are used in an endless array of foods: ice cream, cheese, cereals, candy, salad dressings, soft drinks, and canned fruits, to name a few. Pet food contains dye to make it more appealing to people, not to dogs and cats, which are color-blind.

In the early 1970s, there was controversy about Red Dye Number 2; eventually, its use was banned in the United States. The dye was originally derived from a coal by-product called coal tar. The coal tar "family" of dyes has a bad record. Since 1919, about a dozen coal-tar derivative dyes have been banned in the United States. Their proven harmful effects include cancer.

Most food dyes in wide use today are also coal-tar derivatives. By the time a person is twelve years old, he or she has consumed an estimated three pounds of these dyes. The FDA has authority to ban them, but has not completed safety studies and so allows their continued use.

These dyes include Citrus Red Number 2, a suspected carcinogen, but one that isn't ordinarily consumed, since the dye is used to color the skins of Florida oranges. More worrisome are dyes like Blue Number 1, which is added directly to candy, beverages, and baked goods. It has been banned in Britain as a carcinogen. And Orange B dye, used in hot dogs, causes cancer in mice. Consumer groups argue that people should not be exposed to possible health risks for the sake of nicer-looking food. Some food dyes are safe, but since dye names are not presently listed on product labels, if people want to avoid the potentially dangerous ones, they have to avoid all foods with artificial coloring. More detailed food ingredient labels, proposed by the FDA, the Agriculture Department, and the Federal Trade Commission, may make it easier for consumers to figure out the contents of processed food.

Artificial colors and flavors are suspected of causing hyperactivity, a behavioral disorder of children. The main supporter of this idea is Dr. Benjamin Feingold, a physician and former head of the allergy department of a San Francisco hospital. Dr. Feingold has reported success in treating hyperactive children by eliminating synthetic dyes and flavors from their diets. In about half of the cases, according to Dr. Feingold, the children return to normal or near-normal behavior. A diet recommended by Dr. Feingold is now used by about 20,000 families with hyperactive children.

Some psychologists suggest that the increased attention children get while on a special diet may be the real reason for their improved behavior. And one major study found no evidence that an additive-free dict helped hyperactive children. Dr. Feingold charged that the research was biased,

Adding inexpensive and usually unneeded vitamins to so-called junk foods enables manufacturers to call them "fortified."

since it was paid for by the Nutrition Foundation of New York, an organization that he called "100 percent industry." Several other studies had inconclusive results, but they tested daily doses of food dyes that were less than the amounts usually consumed by children. (The FDA found that children consume an average of 75 milligrams of food dyes each day and that many children take in twice that amount.) In 1980, two studies based on these daily amounts of dyes were reported in the journal *Science*; the results strengthened the evidence that food dycs can provoke disturbed behavior in some children.

Hyperactive behavior has also been blamed on a biochemical abnormality—difficulty in absorbing sugar into the blood—which can be treated by a special diet. This theory is in dispute, too. More research is needed to establish to what extent certain foods, food dyes, and artificial flavors play a part in hyperactivity.

Even if a strong case is made against synthetic dyes, this would be no guarantee that these food additives would be banned. This is especially true when an additive has some proven benefits, as is the case with nitrites, an additive to bacon and other meats.

Sodium nitrites were first added to meat, poultry, and fish because the chemicals react with bacteria to produce an attractive red or pink color. Then a real benefit was discovered: Nitrites retard the growth of botulinus spores, which can cause the sometimes-fatal disease botulism. Adding nitrites to foods has helped (along with refrigeration and freezing) reduce the risk of botulism poisoning to almost zero in the United States. However, medical records show that meats were never a major source of botulism before the use of nitrites. Almost all cases of botulism poisoning then were the result of improper home canning of vegetables.

In the 1920s, the FDA put nitrites on the GRAS list although no safety tests were done. Nitrite limits for cured meats were set at 200 parts per million, simply because that was the usual amount found in cured hams then. The limit was still the same in 1980.

In the early 1960s, it was discovered that nitrites can combine with other chemicals in foods or in the mouth or stomach to form nitrosamines. The process is called nitrosation. And nitrosamines were found to be among the most powerful carcinogens known. They seem to be most effective at causing cancer in laboratory mammals when consumed in small doses over a long period—the way people have been exposed to them.

In 1971, research sponsored by the USDA revealed that nitrites are converted to nitrosamines in food even before it is eaten. A level of 10 parts per billion of nitrosamines in food is considered a cancer hazard to humans. Amounts found in hot dogs were up to 80 ppb, in bacon up to 108 ppb, and in bacon drippings up to 207 ppb. Nitrite-cured bacon has been called "probably the most dangerous food in the supermarket."

As evidence against nitrites grew, so did pressure for restrictions or a ban on their use. The powerful meat industry resisted, and government agencies approached the problem gingerly. A consumer lawsuit, not government action, was instrumental in causing two baby-food companies to stop using nitrites. And when a federal ban on nitrites in baby foods was finally proposed, it came four years after such a ban had been proposed for *pet foods*.

The meat industry gradually reduced the amount of nitrites in their products. The USDA also recommended that vitamin C (or another, similar chemical) be added to cured bacon. Research shows that these substances interfere with the nitrosation process in the stomach. High levels of vitamin C are found naturally in fruits and in lettuce and other green leafy vegetables which contain nitrites. (The incidence of stomach cancer in the United States has declined during the past forty years, perhaps in part because of increased vitamin C consumption over the same period.)

In 1978, a new study seemed to show that nitrites themselves caused cancer in the lymphatic system of rats. However, a group of government and independent pathologists reviewed the evidence and found that the study was inconclusive. The USDA and the FDA continued to try to persuade the meat industry to reduce the use of nitrites in foods. The American Meat Institute, the main lobbying organization of the industry, continued to question the validity of the evidence against nitrites, and to emphasize their value in preventing botulism. In 1980, the USDA and the FDA asked the National Academy of Sciences to evaluate all of the scientific evidence against nitrites and also to evaluate possible substitutes for nitrites.

There are several alternatives to the use of nitrites. Canned meats are sterilized with heat and need no preservatives; nitrites have been added only for color. With the exception of bacon, other meats and baby foods can be precooked too. Freezing or freeze-drying would also minimize the danger of botulism poisoning.

The benefits of a nitrite ban are unmeasurable, though a further decline in the incidence of stomach cancer might be attributed to it. But consumer advocates say that proof of benefits is not necessary. The proof of potential harm to humans is well established. Compared to other health hazards, harmful food additives are easy to control: Just stop adding them. Halting their use should be the first line of defense in achieving a more healthful food supply.

Some drugs and medicines also have no proven benefits, and even some with benefits turn out to be dangerous in other ways. The laws regulating drugs have loopholes, and the enforcement effort by the FDA has been spotty, at best. Too often, a drug is marketed for years and used by millions of people before its harmfulness is discovered.

Only in 1979 was it learned that the drug reserpine caused cancer in mice. Reserpine had been in use for twenty-five years by millions of people. It was widely used to control high blood pressure. Another chemical compound, methapyrilene, was found to be a potent carcinogen in rats. It had been used in nonprescription sleeping pills, nasal sprays, and decongestants—over all in about thirty-six products used by 10 million Americans.

More bad news like this lies ahead, not because "living causes cancer," but because the pace of testing chemicals has increased. At first, the National Cancer Institute (NCI) testing program focused on pesticides and herbicides but then concentrated on a variety of chemicals, especially those to which large numbers of people were exposed. It is important to know that most of the chemicals tested by the NCI —about a hundred a year—have *not* been found to cause cancer or other disease.

Although the bad news does at time seem overwhelming, people can take positive steps to reduce the health risk in their daily lives. The single most important step is to change diets and, in some cases, ways of preparing foods. This may call for a turning away from old family customs and national traditions, including steak or other fatty meat cooked close to charcoal (which results in deposits on the meat of a carcinogen called benzopyrene) and meat cooked directly on a hot metal surface (shown to change sub-

Eating less fatty meat and more foods rich in starch and fibers strikes a blow for good nutrition.

stances in the meat so that it was mutagenic in the Ames test). Cooking lean meat in a broiler, with the heat source *above* the meat, is safer.

Adopting a diet low in fat, sugar, and salt is not as difficult or unappetizing as it may seem. (You will find further information in books and periodicals listed at the back of the book.) And, although the whole complexity of cancer formation is far from understood, we do know that some food substances seem to inhibit that process. These include the additive BHA (a preservative added to cereals and baked goods), and substances called indoles, found naturally in cabbage, broccoli, turnips, Brussels sprouts, and cauliflower.

Vegetables like these don't taste at all like a fast-food ground-meat burger which comes laced with fat and loaded with 1,500 milligrams of salt, but they may eventually be a key part of a new and much healthier American diet.

CHAPTER SIX

Other Hazards in Our Environment

A change of diet and stricter control of the chemicals added directly to foods can greatly reduce daily risk to health. But some dangerous substances are not so easily avoided.

Aflatoxin is a potent carcinogen which results from improper drying of grains. It is a naturally occurring substance produced by two common molds which grow on moist grain and nuts. In laboratory tests, aflatoxin has caused liver cancer in several kinds of mammals. It is the most powerful carcinogen ever tested on rats.

Its effects are most prevalent in tropical countries where moisture and poor storage conditions encourage the growth of mold. Epidemiological evidence from such nations as Kenya, Mozambique, and Thailand show a high intake of aflatoxin and high rates of liver cancer.

This dangerous substance was first identified in 1961.

Accidental chemical contamination of grains that are fed to chickens and livestock may reach the animals at the end of the food chain: humans.

Since then, in the United States, aflatoxin has been detected in several kinds of grains, peanuts, tree nuts, and in milk. Aflatoxin gets into milk when cows are fed grain that is contaminated with this poison.

The FDA classifies aflatoxin as an unavoidable contaminant. The danger varies from year to year, depending on weather conditions, insect populations, and other factors which damage crops. But a lot of contamination is avoidable by proper harvesting, drying, and storage. The yellow greenish mold which produces aflatoxin is easily seen, and the infected food can then be discarded. Some kinds of

Aflatoxin-producing mold on an ear of corn

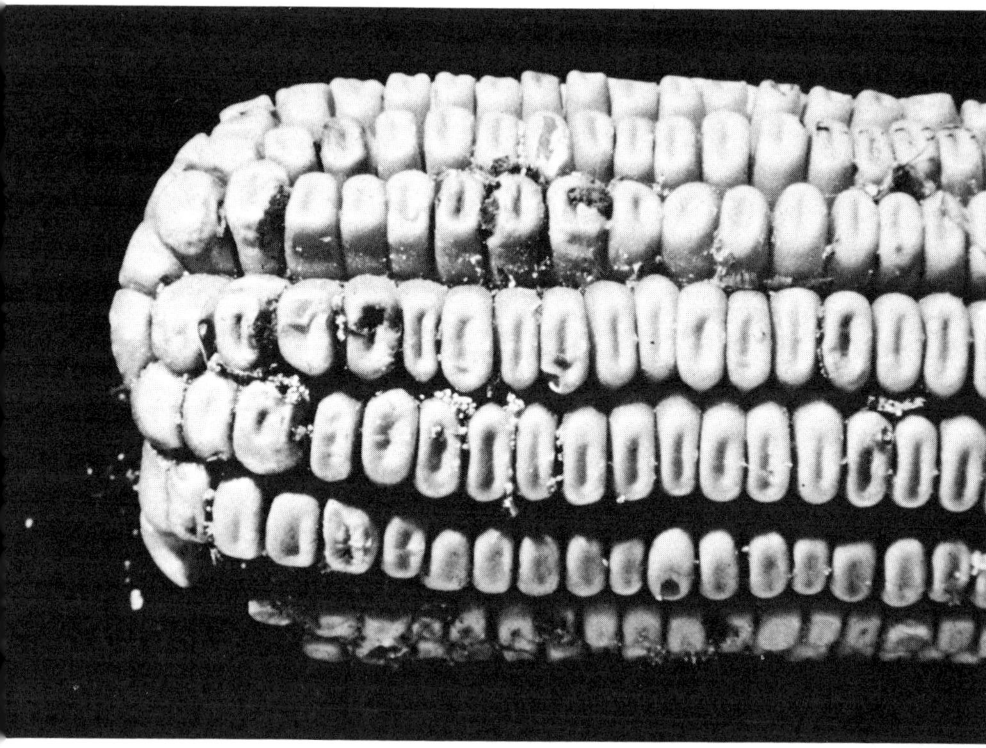

foods can be salvaged. In the usual process of refining peanut oil, for example, any aflatoxin present in the peanuts is removed.

The FDA has set allowable limits of aflatoxin in foods, but consumer groups believe that these standards are not strict enough. Besides urging the FDA to set tougher limits and to enforce existing standards, people can further protect themselves by storing nuts in cool, dry places, and discarding any nuts that are discolored or moldy.

Aflatoxin is just one of several substances which may contaminate our food, and it is more easily controlled than others. A 1979 study by the General Accounting Office, the investigative agency of Congress, found that 14 percent of the fresh meat sold in supermarkets contained residues of chemicals suspected of causing cancer, birth defects, or cell mutations.

The GAO's findings contrasted with those of the USDA, which found chemical residues in only 2 percent of the meat they sampled. Unfortunately, the GAO's study was more thorough. The USDA tested each meat sample for just one specific chemical or class of chemicals and also checked for just forty-three different drugs and pesticides. In contrast, the GAO checked each meat sample for many chemicals and tested for 143 different drugs and pesticides.

These tests were made on meat collected from slaughterhouses. Usually, from six to twenty-five days are needed to make chemical analysis of meat samples. Meat is shipped within a day of slaughter. Thus, there is no practical way to detect illegal residues of chemicals in slaughterhouses and halt the sale of the meat to people.

The only comforting information in this GAO report was

that the residues were in some cases below the limits set by the FDA. But consumer groups and some health experts believe that the FDA's standards are often too lenient. The controversy over these limits and the troublesome nature of some food contaminants are well illustrated by the problem of PCBs.

Polychlorinated biphenyls (PCBs) are a family of oily, synthetic chemicals which had been used for about fifty years before they became a matter of national concern in 1975. Large-scale production began about 1930. PCBs are among the most stable compounds known and have characteristics that make them highly useful in electrical products. They do not burn or corrode easily and are excellent insulators. PCBs were widely used in transformers, condensers, and capacitators (such as those in air conditioners, refrigerators, television sets, and fluorescent light boosters). They were also used in a variety of other products: lubricating oils, paints, pesticides, tape and envelope adhesives, plastics, and upholstery fabrics.

Until recently, millions of pounds of PCBs contaminated the environment each year. The chemicals were accidentally spilled, leaked, or vaporized into the atmosphere. PCBs are often recovered and reused from large electrical equipment but not from other products. When wastes are burned in incinerators or buried in landfills, PCBs reach the air, soil, or water. Some industries have deliberately discharged PCBs into rivers.

There are only a few methods known for destroying PCBs, and none is likely to occur naturally. This means that the PCBs in the environment will be around for a long time. Like DDT, they tend to accumulate in the fat of animals

By the time meat samples from slaughterhouses are tested for chemical residues, the remaining meat has been sold to people.

and to become more concentrated in animals at the top of food chains. Large fish, sea birds, and sea mammals have higher concentrations of PCBs than the creatures upon which they feed. Swedish scientists learned that high PCB levels affected the reproduction of gray seals of the Baltic Sea and caused a steady decline in seal numbers.

Humans are also large mammals at the top of food chains. Evidence of harm to people by PCBs has been traced back to the 1930s. According to old medical reports, workers who handled PCBs developed severe acne on their faces and bodies, and also suffered from fatigue and loss of appetite. Some died due to liver damage. A few warnings about the dangers of PCBs appeared in medical journals, but drew little notice.

An accident in Japan finally focused attention on these dangers. In 1968, PCBs leaked from a heating system into rice oil that was later consumed. More than a thousand people were poisoned. The victims suffered from fatigue, nausea, swelling of the limbs, and liver injury. Babies born to women victims were abnormally small and developed poorly.

The danger of PCBs varies and depends partly on the age and the species of animal afflicted. Mink are extremely vulnerable, while some other animals can tolerate high doses. Older fish are much better able than young ones to live with several parts per million of PCBs in their fat tissues. Young animals, in general, seem vulnerable to harm by PCBs. This fact worries health officials because human babies are exposed to these chemicals both before and after birth.

PCBs are stored in fat, and since human milk has a high fat content, PCBs have been found in milk consumed by breast-fed babies. A 1976 survey of forty-six states by the EPA found PCBs in the milk of one-third of nursing mothers. No one urged that the practice of breast-feeding be stopped. However, health officials urged that pregnant and nursing women eat little or no freshwater fish, the main source of PCBs for most Americans.

Part of the worry about the potential harm of PCBs to human babies results from a study of rhesus monkeys, a species that is relatively sensitive to PCBs. Female monkeys were fed a diet of 2.5 parts per million of PCBs. Their babies were undersized at birth and quite vulnerable to infection. Later on, the young monkeys were hyperactive and had learning difficulties.

People in the United States have an average level of 1.8 parts per million PCBs in their fat, but some individuals have higher concentrations. One nursing mother (who, incidentally, ate a lot of freshwater fish) had more than 10 parts per million in her milk. So far as we know, humans are not as sensitive to damage by PCBs as rhesus monkeys, but some human infants may already be vulnerable to harm from the PCBs in their food chain.

Studies of freshwater fish show that larger fish have more PCBs than small ones of the same kind. Trimming off the skin and fat before cooking reduces PCBs by about half. In some areas, however, even this reduction of the PCBs may not make fish safe to eat.

New York's Hudson River is the best known site of PCB pollution, but hardly the only case. Trouble spots exist in river systems from Massachusetts to Alaska. The Great Lakes are also contaminated. Even Lake Superior, which has little industry near its shores, has PCBs. Scientists estimate that most of the PCBs reach the waters of Superior and the other Great Lakes via rain or snow.

Foods other than fish also contain PCBs. Inspectors have found high levels in eggs, chickens, and turkeys—the result of grain contamination. In 1979, PCBs leaking from an electrical transformer stored in a Montana shed eventually contaminated livestock feed and then human foods in seventeen states. Five million dollars worth of food had to be destroyed. PCBs are also found in food packaging made of

paper. One way or another, low levels of PCBs reach everyone. The danger of these low levels is unkown.

In 1973, the FDA set temporary legal levels of PCBs in foods and food packaging. For poultry and fish, the limit was 5 parts per million. This figure seemed to be picked because it allowed most fish and other PCB-contaminated products to stay on the market. Canada allowed only 2 parts per million, and consumer groups urged the FDA to do likewise. The FDA was slow to act, but a lawsuit by the Environmental Defense Fund finally brought results. In 1979, the FDA lowered the limit of PCBs allowed in fish, poultry, and dairy products to 2 parts per million. The annual loss to commercial and sports fisheries was estimated at anywhere from $6 million to more than $200 million.

Sediments collected from river and lake bottoms (left) are analyzed for deposits of PCBs. Commercial fisheries in Lake Michigan and other areas suffered great economic losses from PBC contamination.

PCBs have been banned from many products but will continue to contaminate the environment for a long time. No recovery program has been set for the 750 million pounds of PCBs in use or for the 290 million pounds in landfills and the 150 million pounds that are circulating in the biosphere of soils, water, and air. It seems inevitable that we will learn more about the harmfulness of PCBs by the least desirable method—using people as guinea pigs.

Another chemical relative of the PCB family, the poison DDT, and several other pesticides are banned for most uses in the United States. Tests show that DDT concentrations are declining in lake waters, as they are in human fat. However, people in the United States sometimes get rather heavy doses of DDT and other banned pesticides from imported food. The FDA found residues of DDT, lindane, dieldrin, and heptachlor (all banned for most uses in the United States) on imported coffee beans. Very high levels of these and other pesticides have been detected on fruits and vegetables from the Caribbean and Central America. And the FDA's commonly-used tests for imported foods cannot detect residues from scores of pesticides, including types that have never been allowed in the United States. One environmentalist, observing this ironic situation, said, "It seems only fair that we eat poisons which are outlawed here but which our State Department and Commerce Department promote in other countries."

The export-import route is a roundabout way of being harmed by the food we eat, but there are other, more com-

Millions of chickens have had to be destroyed upon discovery of contamination by PCBs or pesticides.

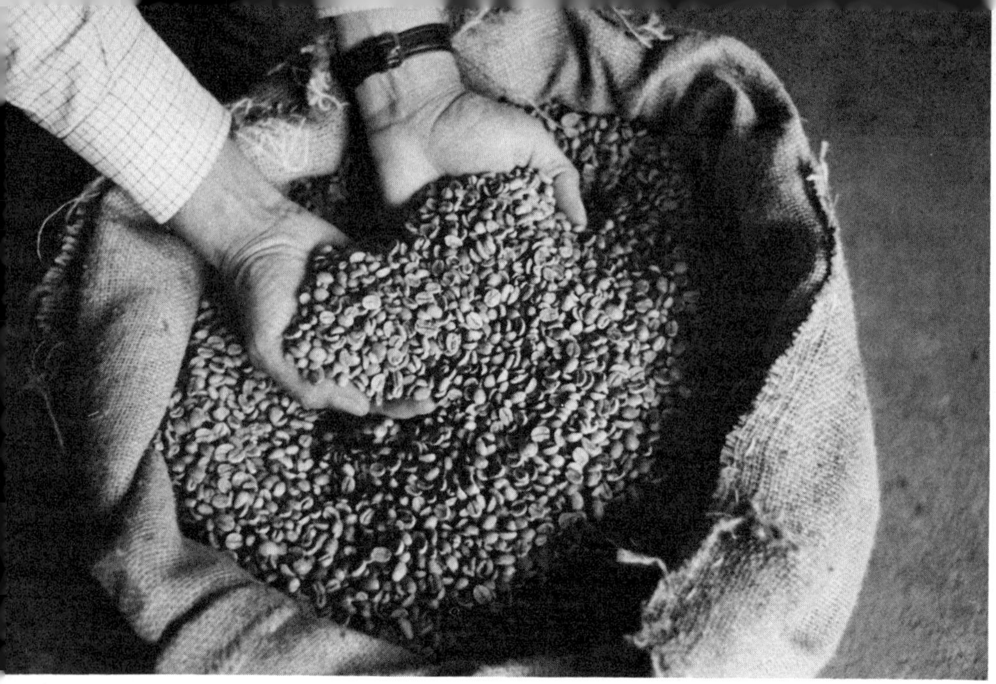

Imported foods, including coffee beans, often have residues of pesticides banned in the United States.

plicated ways. An example is the chemical compound diethylstilbestrol (DES). It is a synthetic hormone which was, until 1979, used in cattle feed. Its first use, however, was as a drug for pregnant women; it was often prescribed to prevent miscarriage.

This drug was given to perhaps 2 million women in the United States between 1940 and 1971. As early as 1953 there was evidence that DES had no value in preventing miscarriages. In 1971, a Boston doctor reported several unusual cases of cancer in the *daughters* of women who had taken DES. Since then other DES-daughters have developed the same cancer or have been unable to have children themselves. The nationwide incidence of this cancer in DES-daughters is low, but it would be difficult to overstate the tragedy of this episode.

The medical use of DES is now limited by the FDA. Until

1979, however, DES had another use and was reaching many more humans by a different route. DES was added to grain fed to cattle and sheep, or implanted under their skin. It stimulated growth, so livestock reached full weight faster —while using less grain—than livestock not given the drug. However, residues were found in meat, and DES later was found to be a carcinogen. In 1972, a consumer protection organization called the Public Citizen Health Research Group began legal action aimed at stopping this use of DES.

Through court suits, the livestock and drug industries successfully delayed a DES ban for seven years. In 1979, their last legal maneuver failed, and the FDA finally stopped the use of DES in livestock feed. The meat industry's forecasts of a resulting rise in beef prices did not occur. However, in 1980 the FDA found that DES was being used illegally at feedlots in twenty states.

DES was just one of many chemicals fed to livestock. Forty percent of all antibiotic drugs made in the United States are added to livestock and poultry feed. Though most of these chemicals seem to have no direct effect on people, their indirect, long-term effects may cause serious outbreaks of disease.

In the 1940s, agricultural scientists discovered that farm animals grew faster when their diet was supplemented with antibiotics. Later, scientists learned that this effect does not occur with healthy, well-kept animals. Antibiotics compensate for poor care and substandard food, so they quickly became a handy crutch for the meat industry.

Cattle and chickens are now fed antibiotics from birth to death. The meat industry claims that antibiotic use is essential; removal of the drugs would cost consumers and farmers up to $2 million a year, they say. This is a "straw man" argument, since no one has suggested removal of *all* feed additives which promote growth. And the meat industry's economic calculations do not take into account the human health costs which may result from continued overuse of antibiotics.

Heavy doses of antibiotics kill all or most bacteria. But low doses, such as those added to livestock feed, stimulate the evolution of strains of bacteria that are resistant to antibiotics. Some of these resistant strains are harmful to humans and some are not. Among the harmful types are *Clostridia*, the cause of tetanus, and *Salmonella*, the cause of paratyphoid and food poisoning. Some resistant bacteria are able to pass this characteristic to other types of bacteria. The overuse of certain antibiotics by doctors has stimulated evolution of resistant bacteria in people, too. Hospitals are now experiencing outbreaks of resistant diseases. Patients have died because doctors discovered too late that the bacteria were resistant to the antibiotic first prescribed.

This unintentional encouragement of bacterial resistance has been going on for years. In the 1960s, the effects of antibiotics fed to cattle were blamed for a disease outbreak in England which caused more than 500 cases of illness and fourteen deaths by paratyphoid. Since then, England and several European nations have restricted or barred certain antibiotics from animal feed. They have since experienced a reduction in sicknesses caused by resistant bacteria.

In 1972, the FDA began an effort to regulate use of penicillin and tetracycline, the two most common antibiotics in animal food. Powerful economic interests, especially the drug industry, blocked action by the FDA and, in 1978, influenced Congress to handicap the FDA's efforts further. Some of the political and scientific issues involved were revealed in a controversy over a report by the Council for Agricultural Science and Technology (CAST).

This is a private organization, governed by the members of twenty-five agricultural trade and research groups that make it up. CAST's alleged aim is to provide impartial information on disputed issues in agriculture. Since 1972, it has

Low doses of antibiotics, routinely fed to livestock, stimulate the evolution of resistant "Super Germ" disease organisms.

published more than seventy reports, most of which tend to favor the status quo and to de-emphasize possible dangers of chemical farming. CAST reports defended use of nitrites in meat and DES in livestock feed.

To the politically astute, these results were wholly predictable, after a glance at CAST's membership and funding. The member societies have long-standing ties with manufacturers of pesticides, drugs, and fertilizers. About two-thirds of CAST's budget comes from these manufacturers. Nevertheless, CAST had a reputation for impartiality, at least among gullible politicians and reporters. In 1979, however, this reputation was damaged in a dispute over CAST's report on antibiotic additives to livestock feed.

None of CAST's members had much expertise on the subjects of bacteria and antibiotics, so seven microbiologists were recruited in 1977 to give the report more credibility. When the microbiologists read the final draft of the report, they promptly resigned, charging that their findings had not been presented accurately.

In their letter of protest, the scientists accused CAST of omitting unfavorable evidence on the risks of adding drugs to livestock feed. Favorable data were emphasized. In general, the evidence was "bent" to fit the public relations aims of CAST.

In 1978, Congress asked the National Academy of Sciences to recommend and conduct a study that would answer, once and for all, the question of whether antibiotics added to livestock feed pose a threat to humans. An NAS committee considered the request and in 1980 concluded that it was technically impossible to find a definitive answer.

It is impossible to trace a resistant bacterium from a sick person back to a farm animal.

Many scientists support a ban on certain drugs in livestock feed, simply because it is known that antibiotics are causing an increase in the world population of resistant bacteria. And the livestock industry has been offered alternatives; two substitute antibiotics rarely used in human medicine have been suggested by the FDA. So far, the livestock industry has rejected this change, because, according to Dr. Stanley Falkow, professor of microbiology and medicine at the University of Washington, drug companies have misled meat producers into thinking that alternative drugs won't be as effective.

In 1976, a *Salmonella* epidemic in the nursery of a Connecticut hospital was traced to infected calves on a farm where the mother of one infant lived. Until antibiotic alternatives are used, resistant bacteria will become more common and the risk to human health will continue to grow.

78-S
79-S
80-S
81-S

CHAPTER SEVEN

Society's Guinea Pigs

In 1972, Olvin Ferguson took a civilian job with the United States Navy putting metal shields around nuclear reactors in submarines. The Navy assured him that the amount of radiation he received—about 3 rems a year—was completely safe. Six years later he had acute leukemia of the lymphatic system and only a few years to live. The Navy claimed that the leukemia was not job related and that it owed him no compensation.

Olvin Ferguson is one of perhaps one-and-a-half-million workers in the United States who are exposed to low level radiation on their jobs. Of these, by far the largest and most vulnerable groups are the medical professions: doctors, nurses, veterinarians, and medical assistants. Other workers exposed to low-level radiation include industrial radiographers, uranium miners, nuclear plant workers, and scientists who use radioactive materials in their research.

Nuclear submarine shipyard workers are exposed to at least three known cancer-causing substances on the job.

Higher than normal incidence of certain cancers has been found among workers at a government center for nuclear research and waste storage in the state of Washington and among Navy shipyard workers. In the shipyards, however, it is difficult to pinpoint the cause of the cancers because workers are also exposed to such carcinogens as benzene and asbestos.

If people are engaged in a war with health hazards, workers are in the front lines. They are the shock troops who usually suffer in the first attack of a dangerous chemical. Their work often exposes them daily to higher levels of hazardous materials than those that reach the general public. Erik Eckholm has written, "Much of what is known about poisonous or cancer-causing materials has been learned by studying unusual patterns of disease and death among workers—humans, rather than rodents, have been society's guinea pigs."

Some diseases are rare or nonexistent among the general public; you really have to work for them. These include angiosacoma, a liver cancer that occurs among workers exposed to vinyl chloride in the plastics industry, and the bone cancer which appeared among employees who painted radioactive radium on dials of watches and clocks.

In 1977, the National Institute of Occupational Safety and Health reported the results of a nationwide survey: One out of four workers in the United States is exposed on the job to some substance believed to be capable of causing death or disease. The survey also claimed that fewer than 5

Certain jobs expose workers to such hazardous substances as vinyl chloride (top) and roofing materials (bottom).

205

NO
PARKING
ALLOWED

percent of workplaces have industrial health services or active programs to help prevent or reduce exposure of employees to hazards. In many cases, neither the employer nor the employees were aware of the potential dangers in their workplace.

For many years it was generally accepted that 1 to 5 percent of all cancer cases in the United States were job related.

Few employees are tested for early signs of work-related illness. Health services in most industries are woefully inadequate or nonexistent.

In 1978, a report by several federal agencies estimated that the figure should be 20 percent. The report was attacked by industry spokesmen and even criticized by some nonindustry scientists. But a growing consensus seemed to be that the estimate of 20 percent was much closer to reality than the earlier assessment. In 1979, the new estimate was supported by a study released by the American Industrial Research Council, an industry group. Conducted by Dr. Reuel Stallones, dean of the School of Public Health at the University of Texas, this study concluded that the 20 percent estimate was reasonable and might even be too low. Dr. Stallones called the situation a "public health catastrophe."

We tend to think of hazardous chemicals being mostly in dirty factories, perhaps at some mythical "Belchfire Industries," and the victims being mostly blue-collar workers. There are notable exceptions to this image. A hospital operating room, for example, is usually spotlessly clean, but the doctors and nurses are commonly exposed to such anesthetic gases as halothane, nitrous oxide, and methoxyflurane. They develop cancer and liver and kidney diseases about twice as frequently as hospital employees who are not exposed to these gases.

An artist's studio can also be a dangerous workplace. A host of hazardous substances, including arsenic, asbestos, cadmium, benzene, lead, and toulene are common ingredients in the materials used by professional painters, sculptors, potters, woodworkers, and other artists or craftspeople. Home hobbyists may also be at risk. Most art supplies have only vague warnings on their labels and ingredient lists which only a chemist can interpret.

Many occupational health hazards are brought home

after work. Wives, children, and other relatives of workers have died of cancer as a result of exposure to asbestos fibers carried home on clothes, shoes, and hair. The children of lead workers often have dangerous levels of lead in their blood.

Not only do occupational hazards reach into homes, they reach future generations, or the unborn who might be part of future generations. Chemicals in the workplace can affect reproduction at many levels, damaging sex cells before conception and fetuses during development. The human fetus is acutely sensitive to harm from substances its mother may inhale, drink, or eat. Carcinogens can cross the placenta to the fetus and produce cancer. One epidemiological study of 400 children who died of cancer before the age of five revealed that an unusually high percentage of their fathers had been exposed to hydrocarbons at their places of work and may have carried these home on their clothing.

Fifteen percent of all babies born in the United States have physical or mental birth defects, mostly from unknown causes. Also, about 30 percent of all pregnancies in the United States end in spontaneous abortions. These problems, already serious, may get worse as more and more women go to work, especially in jobs that used to be almost exclusively occupied by men.

Some industries exclude women of childbearing age from jobs where they would be exposed to certain toxic chemicals. Sometimes, women work at such jobs, then later are forced to transfer to other jobs. Many are fired, or quit, because they are offered a job with lower pay. Companies call this "preventive discrimination." Some women employees and union officials call it real discrimination—against a

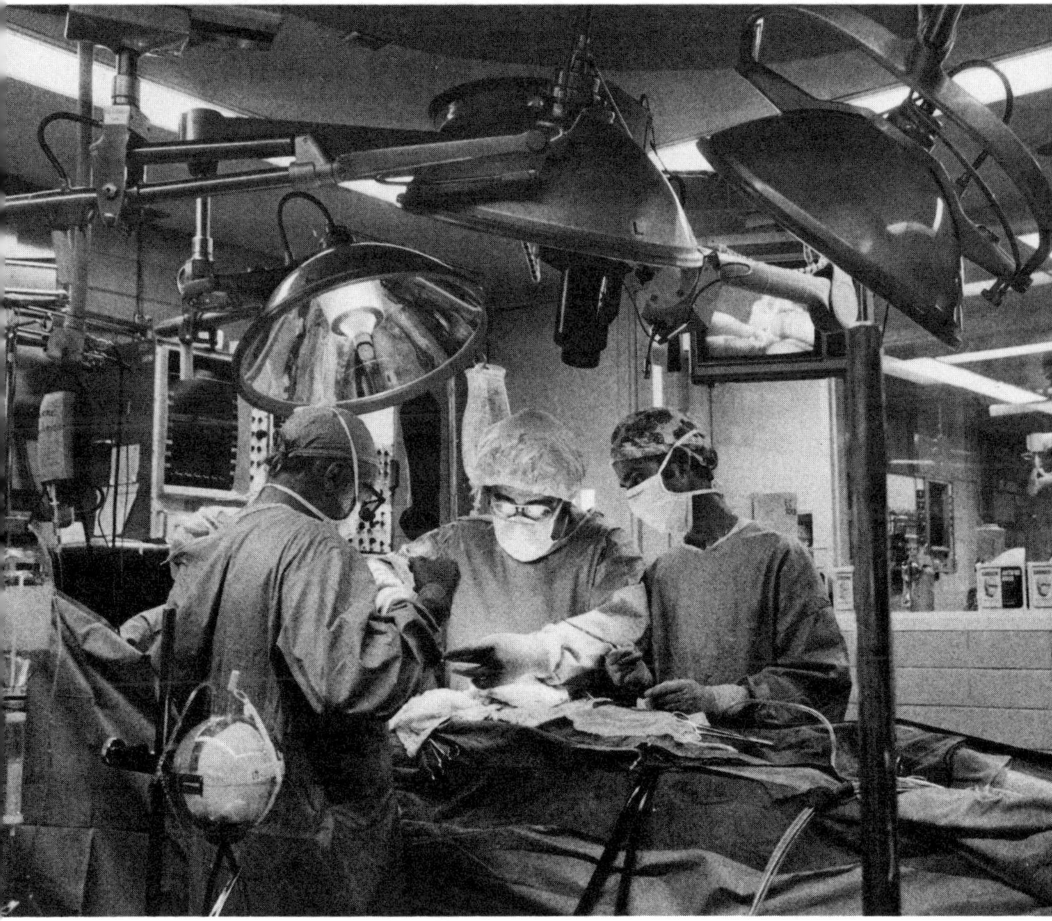

Women who work in hospital operating rooms suffer an increased incidence of miscarriages: so do the wives of male workers.

worker's right to conceive and bear healthy children. They claim that a company's policy of excluding women from certain jobs is an economic choice—to avoid possible lawsuits and to save the expense of cleaning up the workplace. If companies continue to be allowed to remove women of childbearing age from these jobs, one union official said, "We will have established the principle of altering the

The great increase of women in the work force has raised new occupational health issues.

worker to the configuration of the workplace instead of altering the configuration of the workplace to protect the worker."

Men are rarely removed from jobs to protect their reproductive health. Yet they are susceptible to sex cell damage which can lead to sterility, to their wives having a miscarriage, or to birth defects in their children. The entire problem of occupational hazards to workers' reproduction and to the unborn is just emerging as an environmental health issue.

In many industries, the most serious health problems are caused by inhalation of dusts, particles, or vapors. The problem of cotton-dust lung disease among textile workers is an example.

Concentrated in the Southeast, the textile industry employs about 800,000 workers. Of these, an estimated 150,000 are vulnerable to, or already afflicted with, byssinosis or brown lung disease. Some workers seem more susceptible than others. The disease is directly traceable to breathing cotton dust. The first symptom is an occasional feeling of tightness in the chest. With continued exposure, other symptoms follow: reduced breathing capacity, chronic coughing, total disability, death.

Although the amount of dust allowed in the air was set many years ago by the United States Department of Labor, workers still suffered from byssinosis, losing their breathing capacity bit by bit. Complaints from workers and unions had no effect. Then, in 1971, the Occupational Safety and Health Administration (OSHA) was formed in the Department of Labor. This new agency was given the task of protecting workers and the authority to set and enforce standards on the conditions of workplaces.

The agency got off to a shaky start. In 1972, during the Nixon administration, OSHA's first appointed head wrote a then-confidential memo which promised that no controversial new standards, such as those dealing with cotton dust, would be proposed until after the 1972 elections. Thus, improved health for workers ranked behind reelection politics.

In 1979, under union pressure, OSHA proposed tough new limits on cotton dust. At that time, however, the president's economic advisors were examining various government actions with an eye toward curbing inflation. Since the textile industry's costs of implementing more strict dust limits would lead to higher prices for cotton products, OSHA was told to delay action on this matter.

There followed several days of intense debate with the Carter administration. Ray Marshall, Secretary of Labor, argued that the change in dust limits was vital and the costs not excessive. According to Marshall, the changes would cost the textile industry $625 million in new equipment and an $84 million increase in annual expenses. The economists tended to believe the textile industry, which claimed that the initial cost would be $2.8 billion, and that it was technically impossible to reduce cotton dust to the proposed limit.

The administration's economists proposed cutting costs by allowing more cotton dust in the air, and protecting workers by having them wear respirator masks. But respirator masks had been investigated and found inadequate, argued Marshall. The masks were often not worn properly and would be a handicap for workers already short of breath from byssinosis.

President Carter decided to support the strict cotton-dust limits, and OSHA's plan was announced. There was one hitch: The textile industry was given up to four years to meet the new requirements. This plan pleased neither industry nor the unions, and both sides made legal challenges to the new regulations. Late in 1979, the U.S. Court of

Appeals in Washington, D.C. rejected union claims that OSHA's standards were too lax. The court also rejected industry claims that OSHA's plan was too restrictive and expensive.

The controversy over brown lung disease is just one skirmish in a developing war that involves industry, labor, and

Reducing the amount of dust in the air of cotton mills is expected to reduce the incidence of brown lung disease.

government. Labor unions have long claimed that people who sell their labor are entitled to work in safety. For many years though, unions were mostly concerned with job safety of a simpler sort, such as protective devices on machinery. Revelations about cancers, sterility, and other harmful effects from newly-recognized chemical hazards have broadened and increased labor's concern about risks to workers.

There is a growing gap between awareness of potential job hazards and efforts to control them. The National Institute of Occupational Safety and Health (NIOSH) has the task of developing the scientific basis for occupational health standards, and has compiled a list of 28,000 toxic chemicals. NIOSH has recommended worker exposure limits for more than a hundred of these substances, yet OSHA set standards for only twenty-three during its first eight years of existence. At that pace, the agency would have needed a century to better protect workers from all existing hazards, let alone new ones.

In 1977, OSHA's new director, Eula Bingham, announced plans for a new, tougher program on regulating carcinogens in the workplace. OSHA proposed a quicker way of rating the potential harm of chemicals, based on the best available evidence but not waiting for scientific certainty; its plan called for very strict limits on those substances that pose a grave danger of cancer.

Industry reacted sharply with its familiar complaints to almost any attempt at regulation. A spokesman called OSHA's program "scientifically unsupportable, administratively unsound, legally wrong, and economically infeasible." Apparently fearful that OSHA's proposal might be adopted by other federal agencies, more than ninety companies and thirty trade associations quickly raised more than a million dollars to form the American Industrial Health Council. Its first task was to plan legal challenges to OSHA's new regulations, which took effect in 1980.

The outcome of this political battle could have major effects on the way government agencies attempt to protect the health of workers and all citizens. Since they have done a rather poor job so far, a lot is at stake for the public.

Even before this controversy was launched, OSHA acted to set a more strict level of exposure for benzene, lowering it from 10 parts to 1 part per million. This action led to a legal challenge from industry, specifically from the American Petroleum Institute. As is common in such cases, the industry attorneys went "judge shopping" and chose the federal court of appeals in New Orleans, where they hoped for a favorable ruling. In October 1978, the appeals court rejected OSHA's attempt to reduce worker exposure to benzene, because no data on costs and benefits had been presented to justify the lower standard.

Some environmental and health laws, such TSCA and the Consumer Product Safety Act, specifically call for a balancing of benefits and risks. OSHA is not required to do this, although its rules are supposed to be "reasonably necessary and appropriate." OSHA maintained that the new benzene limit was technically possible, that the petroleum industry could afford the changes needed, and that unknown numbers of workers' lives would be saved.

The appeals court observed that the agency ought to be able to make "rough but educated estimates of the extent of benefit expected," and that the benefits should "bear a reasonable relationship to the costs." In a sense this means put-

ting a price tag on a worker's head, a kind of arithmetic that OSHA and labor unions found intolerable and scientifically untenable.

The agency and two national labor organizations appealed the case to the Supreme Court, which ruled on the matter in 1980. In a sharply divided 5–4 decision, the Court rejected OSHA's appeal. Three justices maintained that OSHA had failed to prove the existence of a significant health risk before setting its new benzene rule. Two other justices rejected OSHA's appeal for different reasons. The four dissenting justices argued that the plurality opinion "ignores the plain meaning" of the Occupational Safety and Health Act and places "the burden of medical uncertainty squarely on the shoulders of the American worker."

In essence, the Supreme Court had not faced the deeper issue of whether OSHA must justify the costs of regulation in terms of the number of lives to be saved. When that issue is finally faced by the Court, its decision is expected to have a significant effect on the policies and actions of OSHA, the FDA, and all other government agencies responsible for protecting public health.

Gas chromatographs are used to test for the presence of benzene and other carcinogens in workplaces and in products.

NOT
JUST LITTLE
KIDS!
PLEASE
DON'T
LET ME
DIE

CHAPTER EIGHT

Whose Benefit, Whose Risk?

During the 1960s and 1970s, the environmental and consumer-protection movements emerged, and their efforts were largely responsible for passage of new laws aimed at improving the quality of life. But every hard-won victory for cleaner air and water, for better protection from harmful drugs and unfair business practices, and for fewer poisons in the air that workers breathe had the effect of stiffening the resistance of the business community. New government agencies and new authority for older agencies represented troublesome change and new costs. Businessmen wanted less regulation, not more. A powerful backlash began.

An executive of the General Electric Company gave this grim warning: "In our search for a risk-free society, we are in danger of risking everything: our liberty, our hard-won standard of living, our range of choices, our great and venturesome system of private enterprise."

Young people demonstrate against the involuntary risks to families in the Love Canal area.

And the president of the Chemical Manufacturers Association said, "Already, innovation has been stifled, productivity curtailed, inflation fueled, our ability to compete in foreign markets hampered, and our domestic markets opened to cheaper foreign imports."

The Chemical Manufacturers Association launched an ambitious four-year study to determine its own information on the costs and benefits of chemical regulation, related to TSCA. Since the EPA is required to seek the "least burdensome" means of chemical control, the chemistry association wanted to have ready its own figures, to be better able to influence the EPA and, if necessary, Congress and the White House. Whether the figures would be credible remains to be seen. An EPA study indicates that industry routinely overestimates compliance costs. Generally, economic figures have not been examined as sharply as scientific evidence.

Industry spokesmen continued to use their familiar arguments against proposed regulations. They claimed that there was "no evidence" of a harmful effect with some substances; this usually meant that no one had yet bothered to look or that the evidence was inconclusive. They argued that there are safe thresholds for health hazards, even though the consensus in the scientific community is that no such thresholds exist for most known toxins. They ridiculed tests on rats and mice.

Dr. Michael Jacobson of the Center for Science in the Public Interest, in Washington, D.C., said, "We all know that the food industry spokesmen repeatedly say that certain tests are not valid because massive doses of chemicals or additives are given. But that's what they say in public. In private they agree that such testing is valid."

This charade held little promise of success, so new tactics against regulation were vigorously developed. In a period of inflation and economic troubles for many people, industry's attacks on government regulation and bureaucracy had many listeners. And so the nation witnessed the phenomenon of hundreds of corporate spokesmen giving speeches, writing articles, testifying before committees, lobbying, being quoted in the media and in advertising, and all talking about risks and benefits.

No one was against truly beneficial regulation, they said. But we should be wary of spending enormous sums for very slight improvements in health or overall environmental quality. The technology is available to reduce many health hazards to zero or near it. But to do so would bring economic ruin, they forecast.

Besides, "living is risky." This was a central idea of the antiregulation campaign. Everyday life is hazardous. Dr. Richard Wilson, a physics professor at Harvard University, calculated some of the actions a person could take in order to increase his or her chances of death by one part in a million (this is the same as reducing one's life expectancy by eight minutes). Some of the activities were: smoking 1.4 cigarettes, traveling ten miles by bicycle, eating forty tablespoons of peanut butter, or having one x-ray. His point was that we do many risky things every day, and accept these risks, and ought to be more accepting of other risks as well. After all, among the benefits of modern life is a reduction of many of the health risks our ancestors faced.

This argument was supported by other industrial spokesmen. Peter Hutt, a food industry lawyer, said, "We can all wonder at our naïveté in thinking that eating food would somehow be different from any other aspect of our lives, and would entail no risk whatever." He urged that all life hazards be treated equally. Thus, mountain climbers can be warned of the risks they face, and a package label can warn consumers of the risks inside. The choice is left to the individual.

This reasoning about risks led to a sort of defense of individual freedom. As one industrial defender said, in an apparent reference to saccharin, "Skydivers and other daredevils have their freedom—why can't weight watchers?"

If these arguments seem preposterous, it is because they rely on a faulty premise: that people really expect a risk-free life and that regulatory agencies will destroy entire industries in order to achieve it. This is another "straw man" argument, a phony issue to be set up and knocked down with great show. The real goals of environmental and consumer groups, and the likely consequences if achieved, are much less scary. Referring to his specialty of occupational cancers, Dr. Irving Selikoff said, "We are talking about cleaning up industries, not destroying them. We are talking about safety devices and practices which will reduce the level of exposure for many, eliminate exposure for some; about redesigning plants, as present ones become obsolete, so that they are less hazardous in terms of cancer-causing agents; about employing technicians in industry who can identify 'high risk' groups and direct them into pre-clinical treatment programs before a cancerous mass appears. It seems to me that these steps—fully within the capabilities of our industrial engineers—which we can take and can afford, will reduce the amount of cancer and will add only minimally to production costs."

Besides exaggerating the cost consequences, industry officials tended to overlook or downplay the benefits of regulation. Joseph Califano said, "It is, in my judgment, myopic to argue that programs to protect workers are inflationary—if we do not count in our calculations what those programs buy in safety, health, and often greater productivity."

Cancer, heart disease, and respiratory illnesses now cost an estimated $68 billion annually for medical treatment, loss of worker productivity, and loss of earnings. Presumably, the cost would be even higher without existing regulations. To whatever extent tougher health standards can reduce this toll, the amount must be counted as a benefit to the public. The terrible death toll in the asbestos industry demonstrates that the cost of some preventive regulation can be much less than the cost of a *lack* of regulation. It seems reasonable to ask whether society can afford not to regulate business more vigorously than before. In view of the power, objectives, and past behavior of industry and business, there is no reason to believe that the public would benefit from an easing of regulation of health hazards.

In the present situation, businesses enjoy the economic benefits of their activities and pass along risks and hidden costs to workers, consumers, and to society in general. If a truly fair risk-benefit assessment could be made, businesses would be faced with some costs that they have let others pay. For example, they would pay for the job-related illness that wholly or partly disables 2 million workers each year. As matters stand now, only 3 percent of these workers get compensation benefits from their employers. The rest must rely on Social Security or welfare funds (paid by all taxpayers), or on their personal funds and savings.

Consumer freedom of choice, advocated by some industry spokesmen, is not quite as simple as putting labels on products. Some people can't read, or haven't time to bother, or aren't well informed enough to judge what they read. Is a poorly-informed choice really a free choice? Besides, many of today's environmental health hazards are not taken voluntarily. People may choose such risks as auto driving and smoking, but they do not normally choose to breathe asbestos fibers or to have a side order of PCBs at the fish fry.

There are two kinds of freedom: freedom *to* do or not do something and freedom *from* threats which people may not know about, or understand, or against which they may not be able to protect themselves. Health and safety regulations

People choose some risky activities, knowing something about the risks. This is not true of some other health hazards.

are aimed at giving people more "freedom from" involuntary risks.

There are other benefits to regulations. Strict regulations have been shown to stimulate innovation in industry. The failure to ban or restrict the use of a harmful substance may delay the search for a safe substitute. If Congress had allowed the FDA to ban saccharin in 1977, or set a short deadline for its eventual banning, a safe substitute would probably have been developed much more rapidly.

Often overlooked in the controversy over risk-benefit assessments is that such calculations have gone on through-

out history, and still occur, for example, whenever a major change is made, or a new device or product is marketed. However, people making the decision do not often acknowledge that human health and lives may be part of the equation. Even in the case of a war, the choices have not been well-informed, and the choices have never been out in the open. Some people foresee destructive effects on the morale and strength of a society that formally and consciously begins to make such choices, that places a price on a human life, or values one life over another. This sort of bargaining has already been suggested by a nuclear engineer, who argued that saving the lives of a few workers from radiation was unjustified since the money could be better spent reducing some other hazards.

To this reasoning, Dr. Joseph Rotblat of the University of London responded, "Society does not attach the same price tag to every activity. Often enormous sums are spent to combat a rare disease, while more prevalent diseases are neglected. I believe that this emphasizes the instinctive abhorrence of society to express the value of human life in terms of dollars or pounds sterling. Every human life is invaluable, and the fact that certain hazards, like crossing the road or smoking cigarettes, claim numerous lives is no justification for allowing another hazard to continue, if something can be done about it."

Dr. Rotblat added, "All this brings to my mind Oscar Wilde's definition of a cynic: 'A man who knows the price of everything and the value of nothing.' "

Other observers also had strong objections to official risk-benefit calculations. To engage in such deliberations with industry, in the view of Sheldon Samuels, a union official from Washington, D.C., is "like playing Monopoly with the Mafia, which already owns Boardwalk."

Setting up an official risk-benefit evaluation process in government is itself risky; it will probably guarantee an increase in political pressure on government agencies.

An illustration of this was given by Anita Johnson, a former attorney with the Environmental Defense Fund and an expert on food safety regulations. About the Food and Drug Administration she said, "FDA is not a scholarly institution. It is a bureaucratic institution under siege by enormously sophisticated lawyers and lobbyists who know how to generate and present evidence favorable to their products, who hire public relations firms to get their views out to the public, and who know how to use congressmen and other influential people to pressure FDA. In this atmosphere, when human risk is suspected but not proved, even the most honest FDA official might be expected to succumb to industry's arguments that the benefits override the risks."

Dr. Donald Kennedy, while Commissioner of the FDA, said, "I do not want the power to weigh economic benefits against health risks unless Congress explicitly tells me the value of a life."

Scientific uncertainty exists and is exploited by industry. In risk-benefit assessment there is a danger that some scientific knowledge may be overlooked. Typically, a government agency such as the EPA makes decisions about potential health hazards one at a time. Chemicals are also tested one at a time. And arguments about the risks and benefits of regulation are focused on that one substance. But chemicals are not experienced in the environment one at a time. Dr. David Rall of the National Institute for Environmental Health Ser-

Public interest organizations try to counter the lobbying and advertising efforts of powerful industries.

vices has pointed out that chemicals may act synergistically and some substances are cancer promoters. Suppose a mouse is not harmed at low doses of a certain chemical. Does this mean that humans also will not be harmed by low doses? Not necessarily, because a laboratory mouse, unlike a person, doesn't smoke, doesn't take medicines, doesn't drink alcohol, and doesn't eat bacon. Regulators need to keep this in mind when considering the carcinogenic risks of a single substance.

Would official government risk-benefit analysis lead to increased health and safety for people? If the process is open to the full political treatment that can be mustered by industry, corporate health would probably come before public health. A pattern of half-measures, go slow, and no action would continue and probably worsen, with a decline in people's freedom from health dangers. On the other hand, it is conceivable that a fully open risk-benefit assessment, with strong public participation, might be an improvement over the present behind the scenes politicking.

People can take steps to protect their health: adjusting their diet, avoiding certain foods, and shunning other known hazards. We should be prepared for more bad news, partly because several known chemical carcinogens were produced in great amounts only since the early 1960s. They may account for a 10 percent increase in cancer incidence between 1970 and 1976, and may cause further increase in future cancer rates. Also, scientists are playing a grim game of catch-up, testing more and more of the human-made substances which have been part of our environment for years.

Some of them will prove to be harmful. However, they will only be a small fraction of all chemicals in our environment. And nearly all known hazards can be eliminated or greatly reduced. Action, not apathy, will bar them from our lives. So far, few people have chosen to be part of the political process which decides so many issues of environmental health. Government agencies have the legal responsibility to reduce the health hazards in our lives, yet, as examples in this book show, often the agencies must be pushed toward meeting their responsibilities by citizen groups acting in the public interest. (Some of these groups are listed at the end of this book.)

In crucial matters of environmental health, the key question is not whether there will be health risks, or even what the risks will be, but who shall decide what is done about them. We can be sure that elected officials and staff members of government agencies will hear the view of powerful forces opposed to change. It remains to be seen whether they will hear from citizens, speaking up about the changes they want for a healthier environment.

Working for a Healthier Environment

In matters affecting the quality of the air and water, of our food, of workplaces, and of the everyday environment of our lives, the views of ordinary citizens may be virtually unknown to the officials and legislators who have the power to make changes. Few people write to their representatives in government, even though letters and calls from voters can affect legislation. Often the main task of watching out for the public interest falls to small, nonprofit groups which seek to defend or strengthen laws and to goad government agencies into meeting their responsibilities. Several of these national groups are listed below; branches or similar organizations can be found in some states and communities. Supporting their efforts is one of the most effective ways people can help achieve a healthier environment for everyone.

Center for Science in the Public Interest, 1755 S Street, N.W., Washington, D.C. 20009

Community Nutrition Institute, 1146 19th Street, N.W., Washington, D.C. 20036

Environmental Action Foundation, 724 Dupont Circle Building, Washington, D.C. 20036

Environmental Defense Fund, 475 Park Avenue South, New York, New York 10016

Public Citizen Health Research Group, 2000 P Street, N.W., Washington, D.C. 20036

Natural Resources Defense Council, 1725 I Street, N.W., Washington, D.C. 20036

GLOSSARY

ANTIBIOTIC—a natural substance, produced by certain fungi or bacteria, which destroys disease-causing bacteria or inhibits their growth. The first safe antibiotic developed was penicillin, in 1946.

BACTERIA—one-celled microorganisms. Bacteria play a vital role in the decay of dead plants and animals. Most are beneficial, but bacteria are the cause of such human diseases as tuberculosis, diphtheria, gonorrhea, and typhoid fever.

CALORIE—a unit of heat energy. The calorie used to measure energy in our food is the large calorie, or kilocalorie, which is equal to the amount of heat needed to raise the temperature of 1 kilogram of water by 1° Celsius. Each day, adults in the United States need between 2,000 and 3,000 of these calories from their food.

CANCER—a disorder in which body cells grow wildly, producing colonies called neoplasms. Benign neoplasms (tumors) are made up of cells similar to the surrounding tissues and are usually confined to one area. Malignant neoplasms are made up of cells unlike those nearby and tend to spread through the body. Cancer is the second leading cause of death in the United States.

CARCINOGEN—a substance or other factor (such as radioactivity or ultraviolet rays from sunlight) which causes cancer.

CONTROL—in an experiment, the standard of comparison needed to verify the results. For example, suppose some mice are fed a chemical in their food for a year, and a sizable number of them die. This proves nothing about the chemical and cannot even be called an experiment unless the results can be compared to the death rate of another, similar group of mice, the control group, which is kept under identical conditions except that they are not fed the chemical.

DIABETES—a disease caused by a disorder in a person's normal digestion and use of carbohydrates (sugars and starches). It affects the eyes, kidneys, and nervous system. With treatment, a diabetic person can expect a normal life span.

ENVIRONMENT—all of the surroundings of an organism, including other living things, water, soil, climate, and sounds.

EPIDEMIOLOGY—the study of relations between disease-causing agents and communities or populations.

FETUS—in humans, the unborn young from the age of two months to the moment of birth is called a fetus. During the first two months it is called an embryo.

GAMMA RAYS—high-energy radiation emitted by the nuclei of some radioactive elements. Gamma rays, which have great penetrating power, are used to make x-ray photographs.

HEREDITY—the transmission of traits or characteristics from parents to their offspring.

HODGKIN'S DISEASE—an often-fatal disease which resembles cancer and which affects the lymphatic system, especially of males.

HORMONE—a chemical substance within the body that controls its growth, development, or reproduction.

HYDROCARBON—a chemical compound made up of carbon and hydrogen atoms. Benzene and methane are examples.

HYPERTENSION—abnormally high blood pressure in the arteries of a person's circulatory system.

IMMUNIZATION—the process of becoming immune to a disease, usually accomplished by using a weakened form of the disease-causing agent to stimulate resistance to the disease.

LATENT PERIOD—the time during which a disease exists without developing to a stage where it can be easily detected.

LEUKEMIA—a disease of the tissues which produce white blood cells (leukocytes or lymphocytes), causing an excess of abnormal cells in the blood.

LYMPH—a clear fluid that flows through the body in a system separate from the blood circulatory system. Lymph serves to fight infections and to remove proteins and other materials from tissues.

MICROWAVE—radiation having a wavelength between infrared and shortwave radio.

MUTAGEN—a physical or chemical agent that increases

above the normal rate the frequency of mutations in cells.

MUTATION—a change in the genetic material of a cell, which may result in offspring that are somewhat different from the parent.

OVERNUTRITION—the opposite of being undernourished; consumption of so much protein or so many vitamins or other nutrients that health is impaired.

RADIOACTIVITY—behavior of a substance in which nuclei undergo change and emit radiation in the form of alpha particles, beta particles, or gamma rays.

SYNERGISM—the action of two or more factors which has an effect which is greater than the simple addition of their separate effects. For example, two substances acting together may cause cancer but acting alone do not.

TERATOGEN—a chemical or other factor that causes the development of a physical defect in a developing embryo.

X-RAYS—invisible gamma rays, discovered in 1895, which pass through the body and leave a record of its structure on photographic film.

FURTHER READING

AMES, BRUCE N., "Identifying Environmental Chemicals Causing Mutations and Cancer." *Science*, 11 May 1979, pp. 587–593.

ANONYMOUS, *Cancer and the Worker*. New York: New York Academy of Sciences, 1977.

ANONYMOUS, "Drugs: A Consumer Advocate's View." *FDA Consumer*, December 1978–January 1979, pp. 10–15 (an interview with Dr. Sidney Wolfe, director, Public Citizen Health Research Group).

ANONYMOUS, "Everything Doesn't Cause Cancer." A 12-page booklet available from National Institutes of Health, Washington, D.C. (NIH Publication No. 79-2039).

ANONYMOUS, "Safe Drinking Water for All: What You Can Do." Booklet available from the League of Women Voters Education Fund (1730 M Street, N.W., Washington, D.C. 20036).

AREHART-TREICHEL, JOAN, "Chemical Carcinogens: Part of the Problem." *Science News*, June 23, 1979, pp. 411–414.

BARTON, KATHY, "The Other Water Pollution." *Environment*, June 1978, pp. 12–20.

BAZELON, DAVID, "Risk and Responsibility." *Science*, 20 July 1979, pp. 277–280.

BEHR, PETER, "Controlling Chemical Hazards." *Environment*, July–August 1978, pp. 25–29.

BOYLE, ROBERT, and staff members of the Environmental Defense Fund, *Malignant Neglect*. New York: Alfred A. Knopf, 1979.

BREWSTER, LETITIA, and JACOBSON, MICHAEL, *The Changing American Diet*. Washington, D.C.: Center for Science in the Public Interest, 1978.

BUNIN, GRETA, and JACOBSON, MICHAEL, "Does Everything Cause Cancer?" Booklet available from the Center for Science in the Public Interest, 1755 S Street, N.W., Washington, D.C. 20009.

CARTER, LUTHER, "Dispute Over Cancer Risk Quantification." *Science*, 30 March 1979, pp. 1324–1325.

CARTER, LUTHER, "How to Assess Cancer Risks." *Science*, 25 May 1979, pp. 811, 813, 815–816.

DEVORET, RAYMOND, "Bacterial Tests for Potential Carcinogens." *Scientific American*, August 1979, pp. 40–49.

DORIA, JOHN, "Radiation Regulation." *Environment*, December 1979, pp. 25–31.

DOUTT, RICHARD, "Debugging the Pesticide Law." *Environment*, December 1979, pp. 32–35.

ECKHOLM, ERIK, *The Picture Of Health: Environmental Sources of Disease.* New York: W.W. Norton & Company, Inc., 1977.

EPSTEIN, SAMUEL, "Cancer, Inflation, and the Failure to Regulate." *Technology Review*, December 1979–January 1980, pp. 42–53.

EPSTEIN, SAMUEL, "Of Mice and Men: Animal Testing and Human Health." *Environment*, September 1979, pp. 25–32.

EPSTEIN, SAMUEL, *The Politics of Cancer.* New York: Doubleday Anchor Press, 1979 (revised edition).

EPSTEIN, SAMUEL, "Polluted Data." *The Sciences*, July–August 1978, pp. 16–21.

GILBERT, BIL, "All in Favor of Cancer, Say 'Aye'." *Audubon*, March 1979, pp. 54–63.

GOLD, MICHAEL, "Indoor Air Pollution." *Science 80*, March–April 1980, pp. 30–33.

HALL, BOB, "The Brown Lung Controversy." *Columbia Journalism Review*, March–April 1978, pp. 27–35.

HAUSMAN, PATRICIA, and MAYELL, MARK, "Keeping Score for a Healthy Heart." *Nutrition Action*, November 1979, pp. 6–10.

HECHT, ANNABEL, "24,000 Mice Tell Us Much About Cancer Risks." *FDA Consumer*, July–August 1979, pp. 13–15.

HOPKINS, HAROLD, "Nitrates: Focusing on Safety." *FDA Consumer*, February 1978, pp. 9–11.

JACOBSON, MICHAEL, *Eater's Digest: The Consumer's Factbook of Food Additives.* New York: Doubleday Anchor Press, 1976.

JOHNSON, ANITA, "The Case Against Poisoning Our Food." *Environment*, April 1979, pp. 6–13.

JOHNSON, ANITA, "Unnecessary Chemicals." *Environment*, March 1978, pp. 6–11.

MARCUS, STEVEN, "Troubled Waters." *Technology Review*, February 1979, pp. 14–15.

MARSHALL, ELIOT, "Scientists Quit Antibiotics Panel at CAST." *Science*, 23 February 1979, pp. 732–733.

MARX, JEAN, "Tumor Promotors: Carcinogenesis Gets More Complicated." *Science*, 11 August 1978, pp. 515–518.

MAUGH, THOMAS H., II, "Chemcial Carcinogens: The Scientific Basis for Regulation." *Science*, 29 September 1978, pp. 1200–1205.

MAUGH, THOMAS H., II, "Toxic Waste Disposal: A Growing Problem." *Science*, 25 May 1979, pp. 819–823.

NORWOOD, CHRISTOPHER, *At Highest Risk: Environmental Hazards to Young and Unborn Children*. New York: McGraw-Hill Book Company, 1980.

PRINGLE, LAURENCE, *Nuclear Power: From Physics to Politics*. New York: Macmillan Publishing Company, Inc., 1979.

PRINGLE, LAURENCE, *Pests and People: The Search for Sensible Pest Control*. New York: Macmillan Publishing Company, Inc., 1972.

RALOFF, JANET, "Radiation Hazards: The Military Experience." *Science News*, February 11, 1978, pp. 92–93.

RALOFF, JANET, "Radiation Politics." *Science News*, February 10, 1979, pp. 93, 95.

REPACE, JAMES, and LOWERY, ALFRED, "Indoor Air Pollution, Tobacco Smoke, and Public Health." *Science*, 2 May 1980, pp. 464–472.

RODRICKS, JOSEPH, "Aflatoxins: Hazards from Nature." *FDA Consumer*, May 1978, pp. 16–19.

SELIKOFF, IRVING, and LEE, DOUGLAS, *Asbestos and Disease*. New York: Academic Press, 1978.

SMITH, R. JEFFREY, "The Media's Sweet Tooth." *Columbia Journalism Review*, May–June 1977, pp. 28–29.

SMITH, R. JEFFREY, "NCI Bioassays Yield a Trail of Blunders." *Science*, 22 June 1979, pp. 1287–1292.

SMITH, R. JEFFREY, "Toxic Substances: EPA and OSHA are Reluctant Regulators." *Science*, 5 January 1979, pp. 28–32.

SPETH, GUS, "Waiting for a Philosopher King: Making Cancer Policy Until Certainty Arrives." *Sierra*, February–March 1978, pp. 17–19.

SWANSON, JAMES, and KINSBOURNE, MARCEL, "Food Dyes Impair Performance of Hyperactive Children on a Laboratory Learning Test." *Science*, 28 March 1980, pp. 1485–1487.

WARE, LESLIE, "The Pesticide Boomerang." *Audubon*, September 1979, pp. 150–153.

WILLGOOSE, CARL, *Environmental Health: Commitment for Survival*. Philadelphia: W.B. Saunders Co., 1979.

INDEX

Asterisk (*) indiciates photograph

PICTURE CREDITS: American Cancer Society, 53, 72, 113 (both); Dr. Bruce Ames, *from* "Mutation Research" 31:347–364 (1975), by Ames, B. N., McCann, J. and Yamasaki, E., 32; Tom Chalkey, 90; Coca Cola Ltd., Canada, 15 (left); Consumer Product Safety Commission, 35, 124; S. Harris in *The New Yorker*, 130; Herblock Cartoons in *The Washington Post*, 134; Maudlin in *The Chicago Sun Times*, 6; National Institute of Occupational Safety and Health (photographer: Pat Gouduis), 114, 118, 121; Newport News Shipbuilding and Dry Dock Company (Everton), 110; Laurence Pringle, vi, 11, 47, 63, 77, 80, 82, 97, 133 (both), 137; Sean Pringle, 50, 79, 85; Rush-Presbyterian-St. Luke's Medical Center (Chicago), 117; G. D. Searle and Co., 20; Stiefel/Raymond Advertising, Inc., 15 (right); United Nations, 38, 104; U.S. Army Photograph (III-DR-169) (Cpl. McCaughey), 69; U.S. Department of Agriculture Photo, 41 (Soil Conservation Service, by Earl R. Baker), 74 (Marianne Pernold), 92 (Michelle Bogre), 103 (Ray Lustig), 107; U.S. Environmental Protection Agency-Documerica, 22 (both) (Woodson), 43 (Gary Miller), 54 (McCoy), 57 (Danny Lyon), 60 (G. Daniels), 64 (Michael Philip Manheim), 65 (Colonius); U.S. Food and Drug Administration Photo, 5, 8, 27, 94; University of Wisconsin Sea Grant Institute, 100, 101; Wide World Photos, 44, 45, 126.